The Origins of Feedback Control

Otto Mayr

The M.I.T. Press
Cambridge, Massachusetts, and London, England

Originally published by R. Oldenbourg Verlag,
Munich and Vienna, under the title
"Zur Frühgeschichte der Technischen Regelungen"
Copyright © 1969 by R. Oldenbourg, München

English translation
Copyright © 1970 by
The Massachusetts Institute of Technology

Set in IBM Press Roman
Printed and bound in the United States of America by
The Colonial Press Inc., Clinton, Massachusetts

All rights reserved. No part of this book may be reproduced in
any form or by any means, electronic or mechanical, including
photocopying, recording, or by any information storage and
retrieval system, without permission in writing from the publisher.

ISBN 262 13067 X (hardcover)
ISBN 262 63035 4 (paperback)

Library of Congress catalog card number: 72-123250

Preface to the English Edition

The present volume is a slightly revised version of a book that appeared originally in German in 1969. The translation was made by myself, except for that of the various French passages, which was supplied by Joseph Stein of the M.I.T. Press. For advice on Arabic orthography I am indebted to my colleague Dr. Sami Hamarneh of the Smithsonian Institution. Acknowledgment, finally, is due to my wife Louise, who not only typed the manuscript but also gave crucial support to the whole project in her usual sporting and forbearing spirit.

Washington, D.C. Otto Mayr
July 1970

Contents

Introduction	1
I. The Definition of Feedback	2
1. An Example: James Watt's Centrifugal Governor	2
2. The Block Diagram	4
3. Criteria for the Identification of Feedback Control Systems	6
Part One: Antiquity, Middle Ages, and Renaissance	9
II. Feedback Control in Hellenistic Technology	11
1. Ktesibios	11
2. The Oil Lamp of Philon of Byzantium	16
3. Heron of Alexandria	19
4. Ancient Feedback Devices: Summary	25
III. Float Valve Regulators in the Tradition of Ancient Water Clocks	27
1. The Clock of Gaza	28
2. The Clock of Pseudo-Archimedes	28
3. Al-Jazarī	32
4. Riḍwān al-Khurasānī, called Ibn al-Sā'ātī	35
5. Subsequent Fate of the Water Clocks	38

vi   CONTENTS

IV. Float Valves in the Tradition of Heron's *Pneumatica*   40

    1. The Banū Mūsā   40
        a) Float Valve Regulators   42
        b) On-Off Control   44
        c) The Principle of Philon's Oil Lamp   44
    2. Heron's *Pneumatica* after the Banū Mūsā   46

V. Automatic Control in Ancient China   49

    1. The South-Pointing Chariot   49
    2. The Drinking-Straw Regulator   50
    3. Float Valve Regulators   51

Part Two: Modern Times after 1600   53

VI. Temperature Regulators   55

    1. The Thermostatic Furnaces of Cornelis Drebbel   55
    2. Further Temperature Regulators in the 17th Century: Schwenter, Hooke, Becher   65
    3. Réaumur and the Prince de Conti   67
    4. William Henry's "Sentinel Register"   69
    5. Bonnemain   70

VII. Float Valve Regulators   76

    1. The Reappearance of Float Valve Regulators in the 18th Century   76
    2. James Brindley   77
    3. I. I. Polzunov   77
    4. Sutton Thomas Wood   78
    5. Final Acceptance as a Component of Steam Boilers   79

VIII. Pressure Regulators   82

    1. The Safety Valve of Papin   82
    2. Robert Delap   83
    3. Matthew Murray   86
    4. Boulton & Watt   87

IX. Feedback Control on Mills   90

    1. The Mill-Hopper   90
    2. The Fan-Tail   93
    3. Self-regulating Windmill Sails   95
    4. Sensing the Speed of Rotation   99
        a) Centrifugal Fan with Baffle   99
        b) Centrifugal Pendulums   100
    5. Speed Regulation   102
        a) Mead's Regulator   102

	b) Hooper's Regulator	106
6.	The Millwrights: Summary	107

X. The Speed Regulation of the Steam Engine   109

    1. James Watt's Centrifugal Governor   109
    2. The Spread of the Centrifugal Governor   113
    3. The Regulator of the Brothers Périer   115

XI. The *Pendule Sympathique* of Abraham-Louis Breguet   119

XII. Concluding Remarks   125

    1. Completeness   125
    2. The Rise during the 18th Century   127
    3. The General Concept of the Closed Feedback Loop   129
    4. Subsequent Developments   131

Notes   133

Picture Credits   147

Index   149

I. Introduction

Although interest in the discipline of cybernetics is widespread, a historical account of the origins of one of its central subjects, the technology of feedback control,[1] is still lacking. This field is essentially based upon a single idea, that of the feedback loop. The question arises, how this idea penetrated technology, and how it came to take root there.

The following study is characterized not so much by making available new evidence, but by presenting familiar materials from a fresh point of view. First, it attempts to identify all inventions that represent early examples of feedback control; this occasionally will make it necessary to contradict earlier identifications.[2] Second, it focuses on clarifying the inventions' background and on tracing their effects. In the course of doing this it turns out that the concept of feedback, well-defined, as well as subtle, provides an excellent subject for a case study in the intellectual history of technology.

Our subject matter embraces inventions of feedback-control devices in all stages of realization, from the verbally expressed idea to the industrially proven device. Being concerned in the applications of a purely qualitative principle, we have refrained from quantitative investigations (as for example on practical utility, accuracy, or reliability of the devices). Borderline cases have been considered only if historically significant; structures that lack feedback were included only if they had been erroneously described elsewhere as possessing it.

The early history of feedback control must be sought prior to the 19th century. James Watt's centrifugal governor, invented in 1788, was the first feedback device to attract the attention of the whole engineering community and to be internationally accepted. A direct consequence was the drastic rise in the frequency with which feedback control devices were invented. The period to be covered here shall therefore extend from the earliest relevant inventions to the end of the 18th century.

The concept of feedback is abstract; it is not tied to any particular physical medium. In technology it can be used in mechanical, pneumatic, hydraulic, or electrical systems alike. But its main significance today is that it can also be applied profitably in economics, sociology, or biology: The mathematical methods of control dynamics are equally valid in all of these fields. The importance of the concept of feedback is illustrated by the fact that it gave cybernetics its name. When in 1947 Norbert Wiener christened the newly founded discipline (he had not known that in 1834 A.-M. Ampère had proposed *cybernétique* as a term for the science of government[3]), he made use of the Greek word for steersman, κυβερνήτης. He had come upon this through the etymology of the word *governor* (English to *govern*—Latin *gubernare*—Greek κυβερνᾶν), the familiar term for the first popular feedback device.[4]

I. The Definition of Feedback

The purpose of this section is, first, to explain the concept of feedback in terms of an example, verbally and in the language of block diagrams; second, to introduce a few technical terms used in control engineering; and finally, to derive an exact definition of feedback. It addresses itself mainly to readers unfamiliar with control engineering.

1. An Example: James Watt's Centrifugal Governor
Figures 1 and 2 show the steam-engine governor in its earliest form (for historical details see pp. 109ff). Its purpose is to maintain the speed of rotation of the engine (the controlled variable) at a constant predetermined value (command signal) in spite of any changes in load and steam pressure (disturbances). It accomplishes this by sensing the actual speed and adjusting the steam inlet valve of the engine accordingly. The speed of rotation is measured by a pair of centrifugal pendulums. Connected with the flywheel by ropes and pulleys, and rotating at engine speed, they swing outward with increasing speed under the influence of centrifugal force. A linkage in the form of lazy tongs transmits this motion to a sleeve which slides up or down along the axis of rotation of the governor. An arrangement of levers connects the sleeve with the steam valve in such a manner that the flow of steam is throttled with increasing speed.

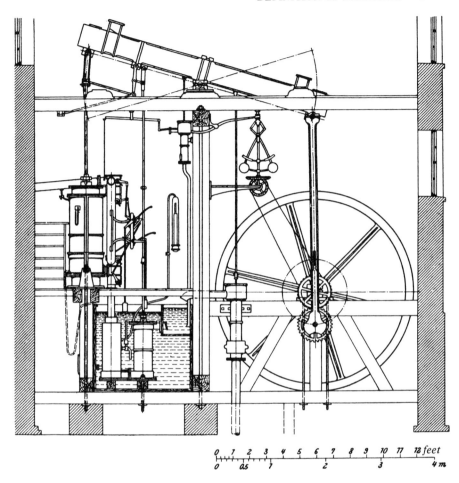

Figure 1 Watt steam engine (1789-1800) with centrifugal governor.

If the load upon the running engine is suddenly increased, its speed will decrease. The flyweights will swing back, and the sleeve will slide upward, causing the steam valve to open. The increase in the flow rate of steam, and hence in torque, will accelerate the engine. The centrifugal weights will fly outward again, reducing the aperture of the valve. Ultimately, the engine will reach an equilibrium at a new speed that lies somewhat below the equilibrium speed prior to the load increase. This *offset* due to lasting disturbances or changes in the command signal is a characteristic of all *proportional* control systems. The increased load requires an increased

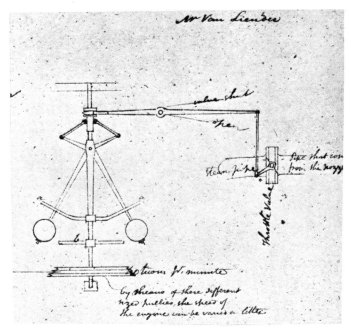

Figure 2   Centrifugal governor, design sketch, Boulton & Watt, 1798.

flow of steam which can be provided only by a changed position of the flyweights.

The desired running speed can be preset by the operator. The command signal is represented by the length of the vertical link on the right of Fig. 2, which connects the large horizontal lever with the small lever on the valve. It could easily be made adjustable by some appropriate device (not shown), as for example a turnbuckle.

2. The Block Diagram

The graphic symbols of the block diagram provide a language in which the components of control systems can be described unambiguously and economically.[5] Indeed, expressing the action of a system in this language is the first step in its mathematical analysis. The notation of block diagrams employs four symbols: The *arrow* (Fig. 3a) designates the *signal*, a quantified physical variable acting only in the direction indicated. Its significance is a causal relationship, not a flow of material or energy. The *block* (Fig. 3b) designates the functional relationship between signals. Signals going into the block are independent variables (inputs, or causes); the signal leaving the block is the dependent variable (output, or effect).

DEFINITION OF FEEDBACK 5

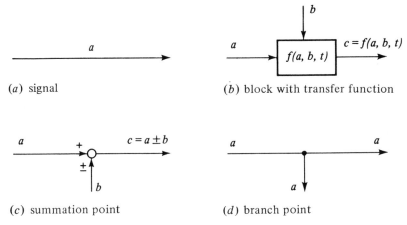

(a) signal

(b) block with transfer function

(c) summation point

(d) branch point

Figure 3 Symbols of block diagrams.

The *transfer function* (based on a notation of linear differential operators) is often written inside the block. The output for a block is then simply the input multiplied by the transfer function. The *circle* (Fig. 3c) signifies a *summation point* or *comparator*. Here two signals are added or subtracted, as indicated by the plus or minus signs written next to the arrows. The *branch point* (Fig. 3d) describes the case when one signal simultaneously causes two separate effects.

Figure 4 shows this notation as applied to the centrifugal governor. It is evident that the sequence of signals forms a closed loop, the main characteristic of all feedback systems. At the left is the input summing point where the command and feedback signals are compared; this results in the *deviation* or actuating signal causing the control action. In this case the disturbances are the steam pressure of the boiler and the load on the engine. If these are assumed to be small in order to linearize the system, they can be described in the diagram as additional summing points. The controlled variable of the system, engine speed, is sensed by the centrifugal pendulum which generates the feedback signal. In this case it is somewhat difficult to identify the elements representing the command and the comparator. The designer of the governor of Fig. 2 thought the desired speed depended on the gear ratio between engine and governor, for he wrote below the three pulleys on the governor axle: "By means of these different sized pullies [sic] the speed of the engine can be varied a little."[6] The block diagram shows, however, that this is not the reference input element. A change of a constant factor (in this case of a transmission ratio)

6   INTRODUCTION

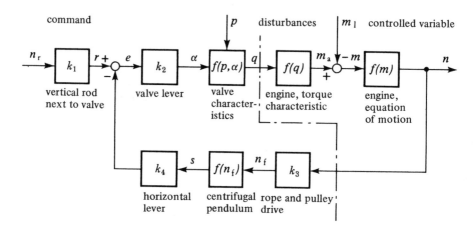

n = desired speed			m_a =	driving torque
r = position of vertical valve rod			m_l =	load torque
b = position of the right end of horizontal lever			m =	$m_a - m_l$
α = $r - b$			n =	actual speed
a = angular position of valve lever			n_f =	speed of governor
p = steam pressure			s =	position of governor sleeve
q = steam flow rate				

Figure 4   Block diagram of the centrifugal governor shown in Fig. 2.

will change only the sensitivity of the system. The command, however, must act additively upon the system through a summing junction. This condition is fulfilled only by the vertical rod above the valve lever. If its length changes, the valve will move, even with the governor sleeve fixed. The comparison between reference input and feedback signal is therefore carried out by subtracting the length of the vertical bar from the position of the right-hand end of the horizontal lever. The result is expressed in the position of the valve. For many feedback control systems the command signal may remain constant during normal operation (such systems are often called *regulating systems*), while on others it may be continually changing. Systems of the latter type, where input and output are mechanical positions, are referred to as *servomechanisms*.

3. Criteria for the Identification of Feedback Control Systems

In 1951 the American Institute of Electrical Engineers published a set of "Proposed Symbols and Terms for Feedback Control Systems" which has since been widely accepted by American engineers.[7] It offers this definition: "A *Feedback Control System* is a control system which tends to maintain a prescribed relationship of one system variable to another by comparing functions of these variables and using the difference as a means of control." The corresponding definition published by the British Standard Institution in 1967 is similar.[8] It defines the term *closed-loop control system* as "a control system possessing monitoring feedback, the deviation signal formed as a result of this feedback being used to control the action of a final control element in such a way as to tend to reduce the deviation to zero." It further specifies *feedback* as "the transmission of a signal from a later to an earlier stage," and *monitoring feedback* as "the feedback of a signal representing the controlled condition along a separate path provided for that purpose, for comparison with a signal representing the command signal to form a signal representing the deviation."

For purposes of the present study, it is essential to define the concept *feedback*, the history of which we are investigating, as rigorously as possible. In order to obtain an instrument with which we can irrefutably identify feedback control systems, we shall summarize their various characteristics in the following criteria. (We will here consider only *automatic* feedback systems, in contrast to manual closed-loop control where the functions of comparison and control action are fulfilled by a human operator. Manual feedback control can be recognized in many timeless human activities, but these are not within the scope of the present study.)

1. The purpose of a feedback system is *to maintain a prescribed relationship of one system variable to another.* The system has the task of automatically maintaining some given variable equal to a desired value in spite of external disturbances. In short, its purpose is to carry out a command automatically.

2. This is done *by comparing functions of these variables* (i.e., command and controlled variable) *and using the difference as a means of control.* In the words of Norbert Wiener, "feedback is a method of controlling a system by reinserting into it the results of its past performance."[9] For the purpose of comparison, a function of the controlled variable—the feedback signal—is transmitted from the output side of the system back to the input side. The cause-and-effect chain which constitutes the system is thus closed, forming the characteristic "closed loop." To say that the control is based on the difference between command and controlled variable implies that a subtraction takes place. Once a signal travels around the loop, its sign must be reversed (the system has *negative* feedback). This requirement is essential. A closed loop without the reversal of sign would be unstable; it would be a "vicious circle."

3. The criteria obtained so far are necessary to identify feedback systems but they do not suffice. Numerous systems exist where input and output are maintained in a "prescribed relationship," and where, either by physical reasoning or by mathematical formalism, a closed loop with negative feedback can be identified. Examples are analog computer programs for differential equations, or simple physical systems with self-regulation,[10] such as the water level upstream of a weir, the R-C circuit, or the weather vane. Indeed, all systems in which the denominator of the transfer function consists of a polynomial containing an absolute member can be represented formally, by means of block diagram algebra, as closed loops with negative feedback.[11] To eliminate systems of this sort we shall consider therefore as genuine feedback control systems only arrangements where the comparator, the feedback path, and the sensing element, or at least one of these, can be recognized as physically distinct elements.

The three criteria we have now obtained contain a sufficiently complete definition of the concept. Briefly repeated, they are:

1. The purpose of a feedback control system is to carry out commands; the system maintains the controlled variable equal to the command signal in spite of external disturbances.

2. The system operates as a closed loop with negative feedback.

3. The system includes a sensing element and a comparator, at least one of which can be distinguished as a physically separate element.

Part One
Antiquity, Middle Ages, and Renaissance

II. Feedback Control in Hellenistic Technology

We usually associate the culmination of ancient technology during the Hellenistic period with the names of the three mechanicians Ktesibios, Philon, and Heron. Among their works, we find the earliest feedback devices known, and it can be shown that the tradition of these inventions survived until the Arabic Middle Ages.

1. Ktesibios

Ktesibios lived in Alexandria, had originally been—according to Vitruvius—a barber, and served as a mechanician under the Ptolemies, more specifically probably under King Ptolemy II Philadelphus (285 to 247 B.C.).[12] He is thus presumed to have lived in the first half of the third century B.C. In this dating,[13] as well as in rejecting the hypothesis of a second, younger Ktesibios, we follow A.G. Drachmann.[14] Ktesibios' own writings, which Vitruvius still had read, have not survived. We know of him only through seven ancient secondary sources, the most informative of which is Vitruvius' *De architectura*.[15] In ancient times Ktesibios had been famous. Vitruvius ranked him as an equal to Archimedes.[16] He is credited with having invented the force pump, the water organ, several catapults, and the water clock. Vitruvius' account of the Ktesibian water clocks contains a passage (9:8:4-7), unfortunately vague, which is a description of the earliest feedback device on record (Fig. 5), at least according to Diels' interpretation. As this interpretation has been contested by some notable

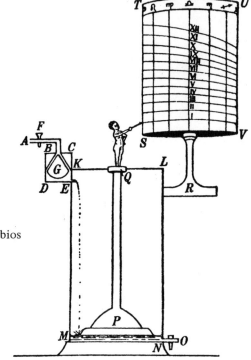

Figure 5 Waterclock of Ktesibios (reconstruction by H. Diels).

scholars, it will have to be examined in some detail, along with the alternatives proposed. The text of the passage reads as follows:[17]

 Primumque constituit cavum ex auro perfectum aut ex gemma terebrata; ea enim nec teruntur percussu aquae nec sordes recipiunt, ut obturentur. 5. Namque aequaliter per id cavum influens aqua sublevat scaphium inversum, quod ab artificibus phellos sive tympanum dicitur. In quo conlocata est regula, versatile tympanum. Denticulis aequalibus sunt perfecta, qui denticuli alius alium inpellentes versationes modicas faciunt et motiones. Item aliae regulae aliaque tympana ad eundem modum dentata una motione coacta versando faciunt effectus varietatesque motionum, (in) quibus moventur sigilla, vertuntur metae, calculi aut ova proiciuntur, bucinae canunt, reliquaque parerga. 6. In his etiam aut in columna aut parastatica horae describuntur, quas sigillum egrediens ab imo virgula significat in diem totum. Quarum brevitates aut crescentias cuneorum adiectus aut exemptus in singulis diebus et mensibus perficere cogit. Praeclusiones aquarum ad temperandum ita sunt constitutae. Metae fiunt duae, una solida, una cava, ex torno ita perfectae, ut alia in aliam inire convenireque possit et eadem regula laxatio earum aut coartatio efficiat aut vehementem aut lenem in ea vasa aquae influentem cursum. Ita his rationibus et machinatione ex aqua componuntur horologiorum ad hibernum usum conlocationes. 7. Sin autem cuneorum adiectionibus et detractionibus correptiones dierum aut crescentiae ex cuneis non probabuntur fieri, quod cunei saepissime vitia faciunt, sic erit explicandum.

First he made an opening of pure gold or of a pierced gem; for these materials are not worn down by the flow of water, nor do they collect dirt whereby they might be obstructed. 5. The water, then, flowing in evenly through the opening, raises up an inverted bowl [a float] known to the artisans by the term "cork," or "drum." Mounted on it is a rule and also a disk that can rotate. Both are equipped with equal teeth which, engaged into each other, carry out corresponding rotations and motions. Similarly, other rods and drums, toothed in the same manner and kept in rotation by one moving force, produce effects and varieties of motions, moving puppets, rotating signposts, letting pebbles or eggs drop, sounding trumpets, and other by-works. 6. Among these also, the hours are marked on a column or pillar, indicated during the whole day by a figurine with a wand, which rises up from the lowest point. Adding or removing wedges for individual days or months takes care of the increases or decreases in the hours' length. Valves to regulate the flow of water are constructed thus: Two cones are made, one solid, one hollow, finished on a lathe in such a way that one can enter and fit into the other and that by the same principle, their loosening or tightening will produce either a strong or a weak current of water flowing into the vessel. Based on such reasoning and mechanisms, arrangements of water clocks are devised for the use in winter. 7. If it should happen, however, that by adding or removing wedges the increases or decreases in the days' length are not accurately indicated by introducing or removing wedges, because these wedges often cause trouble, then one must proceed as follows: ...[18]

Insofar as this text is clear, it presents the following picture of a water clock: Water trickles through a carefully made orifice into a measuring vessel, where the water level rises slowly. With it rises a float, which—via linkages and gears—sets into motion a variety of time-indicating mechanisms. Two points, however, remain obscure. The length of the hours (the "temporary hour" used in antiquity was one twelfth of the interval between sunrise and sundown) is claimed to be adjustable for the time of year by "adding or removing of wedges." Nothing is said, however, about the nature and the location of these wedges. Presumably they were inserted into the orifice itself, restricting the flow of water. As this method admittedly caused difficulties ("quod cunei saepissime vitia faciunt"), an adjustable scale was adopted instead and was calibrated for the different seasons of the year (9:8:7).

The second question concerns the arrangement for supplying the clock with water. If it had consisted of nothing but a simple supply vessel installed upstream of the metering orifice, filled at the start, and then issuing into the clock, then the decrease in the rate of flow caused by the sinking water level would have caused intolerable errors. It was crucial, therefore, that the water should trickle into the measuring vessel at a constant rate. The words *aequaliter... influens aqua*—"the water... flowing in evenly" (9:8:5) prove that Vitruvius and Ktesibios were aware of this necessity. How was the supply vessel constructed that fed the clock with water at a constant rate of flow? The answer obviously lies in the sentences beginning with *Praeclusiones aquarum ad temperandum...—* "valves to regulate the flow of water..." (9:8:6), but they are obscure and have been interpreted in different ways. The wording of the text alone is simply not sufficient for an indisputable reconstruction. We are dealing with a complicated apparatus described inadequately by Vitruvius for lack of either thoroughness or comprehension. Rehm had "... the impression that [Vitruvius] compiled from literary sources with but little understanding."[19] In Vitruvius' defense it must be added, however, that clocks are quite far removed from his main topic, *architecture*.

Rehm suggested the following interpretation for this passage:[20] the word *regula* is taken to mean a valve rod equipped with a thread. By means of this threaded rod the solid cone can be screwed into the hollow cone, thus forming a finely adjustable metering valve. This valve is installed at the outlet of the feed vessel, issuing into a third vessel with overflow, placed between the feed vessel and the measuring vessel. The flow through the metering valve is adjusted by attendants in such a manner that the overflow vessel will always be full. The current into the measuring vessel of the clock is thus assured of a constant head.

The following objections against this interpretation come to mind: Apart from the fact that it is rather farfetched to translate *regula* with "threaded valve rod" (Rehm had used this term not explicitly but only in paraphrase), there is no evidence that such valves were used in antiquity. Nothing in the text refers to an additional overflow vessel, even though Vitruvius would have had no difficulty understanding and describing it. Finally, a clock whose accuracy depended on human attendants would be an invention of questionable value for which Ktesibios would hardly have cared to take credit. Man is singularly unsuited to monotonous tasks requiring long-sustained concentration. Such a clock could never have been more reliable than its attendants.

The hypothesis of the overflow vessel is also supported by Drachmann,[21] who is of the opinion that the wedges were used to adjust the distance between cone and valve seat in order to change the length of the hour; the details of construction or the technical merits of such an arrangement are not discussed.[22]

Water clocks based on overflow vessels have undoubtedly been used in antiquity. They were advantageous when an abundant water supply was available at all times. But to supply a water clock based on overflow from a stationary reservoir would have required a reservoir of disproportionately large size compared with the clock itself. The result would have been an unwieldy arrangement requiring not only much room but also a great deal of attendance. The text gives no support to the hypothetical overflow vessel. Furthermore, this interpretation assigns to the wedges and cones a role which is improbable both from the point of view of the actual text and from that of practical mechanics.

The most convincing reconstruction is advanced by Hermann Diels.[23] He recognizes in the solid cone (*meta solida*) a float, swimming in a regulating vessel upstream of the metering orifice, projecting into the conical valve opening which is connected to the water supply, and throttling the inflow of water with rising level (see Fig. 5: G float, E metering orifice, *BCDE* regulating vessel, A water supply). The solid cone is at once float and valve plug, while the hollow cone (*meta cava*) is the valve seat. With sinking level the valve is opened, water flows in, letting the level rise again. The arrangement achieves a constant level in the regulating vessel and hence a constant flow through the metering orifice.

This interpretation is based on the reading of some Arabic texts on water clocks that have made use of the float valve and that profess to be based on ancient sources (see p. 27). Diels' reconstruction in no wise contradicts the text, and it alone makes otherwise obscure words technologically meaningful. The only philological objection against it is that Vitruvius did not refer to the valve cone expressly as a float. The first sentence of 9:8:5 shows, however, that Vitruvius knew of no specific term for "float"; he used various paraphrases instead. The remaining objections by Rehm and Drachmann, both classic philologists, are nonphilological in character.

Rehm points out that on the Arabic clocks the float valve regulator is employed to control the outflow, and not, as on the clock of Ktesibios, the inflow. This he considers a difference of principle and therefore an argument against a historical connection, although in fact the two arrangements have not only the same purpose but also great physical resemblance (compare Fig. 5 with Figs. 18 to 27).[24]

Drachmann's objection that the purpose of the valve is to regulate not the level but the rate of flow overlooks the fact that the flow is regulated indirectly by the water level.[25] He believes therefore that valve and wedges belong together in some unspecified way. As further evidence against Diels' reconstruction, Drachmann points out that Heron of Alexandria, Ktesibios' heir, was not aware of float valves of this type.[26] Actually, Heron's float valve of *Pneumatica* I:20 (Fig. 10) is equivalent in principle to that of Ktesibios; they have identical block diagrams. Dissimilarities in outward appearance are due to differences in application and to accidents in the transmission of the manuscript figures.

16   HELLENISTIC TECHNOLOGY

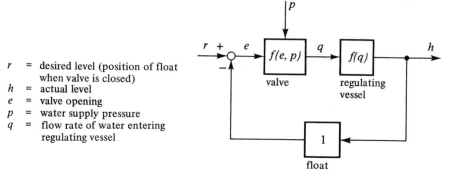

r = desired level (position of float when valve is closed)
h = actual level
e = valve opening
p = water supply pressure
q = flow rate of water entering regulating vessel

Figure 6   Float regulator of Ktesibios, block diagram.

The remaining question is whether the float valve of Ktesibios is a feedback device according to our criteria. The controlled variable of the system, the water level in the small regulating vessel *BCDE*, is sensed by float *G* whose tip is a valve that opens when the level falls and closes as it rises. The command, i.e., the desired water level, is represented by the height of the float, i.e., by the distance between water level and tip. A block diagram (Fig. 6) of the system shows a closed loop.

The only outside disturbance is the pressure in the water supply. The float valve fulfills a dual function: As a float it serves to sense the output, while its upper part in the role of a valve cone performs the control action. The functions of sensing and corrective action are separated, although only indistinctly. Satisfying the criteria established earlier, the float valve of Ktesibios may therefore be regarded as the oldest known feedback device.

In the eyes of his immediate posterity Ktesibios was the outstanding inventor of his era. His intellectual environment was unique. His native city, Alexandria, was the scene of unprecedented scientific achievements that made it the center of the scholarly world. Among his fellow citizens—and perhaps his friends—were men like Aristarchos, Euclid, Archimedes, and Eratosthenes. With such few facts known about Ktesibios, he may well be considered the inventor of the first feedback device.

2. The Oil Lamp of Philon of Byzantium

It is now generally accepted that Philon, whom both Vitruvius and Heron called Philon of Byzantium, lived one generation after Ktesibios, i.e., in the second half of the third century B.C.[27] It is not known where he spent his life; only in passing does he mention having been in Rhodes and in Alexandria. He does not speak of personal acquaintance with Ktesibios, but he hints of having been in touch with persons who had known Ktesibios directly.[28]

Philon is credited with being the author of a 9-part compendium of the mechanical sciences. Only fragments of this have survived, among them a book on pneumatic devices. The *Pneumatica* has been transmitted only by detour via the Arabic, and indeed in two versions, one published in 1902 in French by Carra de Vaux;[29] the other, a fragmentary Latin translation from a lost Arabic copy was translated into German by Schmidt in 1899.[30]

Philon does not describe float valves like those of Ktesibios or Heron, but he knew another method of regulating liquid levels that deserves to be examined here. It is presented in four examples found in chapters 17-20 of the "Arabic" version (Carra de Vaux) or in chapters 12-15 of the "Latin" version (Schmidt). He was more concerned with the basic principle than with solutions of special problems, for he says (this sentence is found only in the Arabic version, following the third of the four examples, Chapter 19):[31]

> These three specimens of apparatus that we have just described belong to the class of constant-level vessels. Employ them as you wish for bathtubs, sinks, or lamps; it is the same thing in all cases.

Further applications can be found in chapters 29, 33, and 34 of the Arabic version. We will illustrate Philon's level control system by the example of the oil lamp (Arabic version, chapter 20, Latin version, chapter 15), the best known of the four basically equivalent systems. In the Latin version it is described as follows (Figs. 7 and 8):[32]

> Similarly make another vessel *abc* which (being a sphere) consists only of one surface. On two sides it shall be provided with the outflow tubes *cd* and *be*, and with a pipe going down vertically into the vessel *ghz* (the belly of the lamp), which is fastened hermetically to both vessels at *l* and *m*. This is the pipe *klmn*. Certain parts of the vessel *ghz* located under the tubes *cd, be*, each under the appropriate one, may be extended on the outside in the fashion of night lamps. These are *gt, sz*. If one fills the vessel *abc* with water below the level *n*, then the liquid will flow through the tube *cd* to *sz* and through the opposite one *be* to *gt* on both sides into the vessel *ghz*, until it reaches the mouth of the pipe *lk* (within the lamp). When this mouth is closed (by the liquid), the outflow at *d* and *e* will stop. Assuming, for example, that the liquid in the vessel *abc* is oil, a wick or paper is placed into the oil of the vessel *ghz*. According to the quantity of oil consumed by burning in *ghz*, oil will gradually flow downward from *abc* through *d* and *e*. This process is of the same kind and has the same meaning.

18   HELLENISTIC TECHNOLOGY

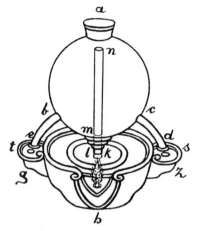 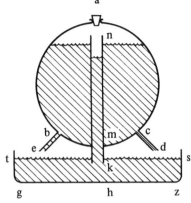

Figure 7 Philon's oil lamp.   Figure 8 Philon's oil lamp, schematic drawing.

In this oil lamp as well as in the three other devices a constant level is maintained by a supply of liquid from the reservoir above. The heart of the arrangement is the vertical riser in the middle. When the oil level has dropped low enough to expose the lower end of the pipe, air will rise through the pipe into the oil container, permitting oil to flow out through the capillary tubes. Consequently the oil level will rise until it reaches the lower end of the riser, stopping the upward flow of air and hence the downward flow of oil.

The proper functioning of this device depends on the shape and dimensions of the tubes, as an experiment by the present author has shown. The riser must be wide enough so that air can rise without resistance as soon as the lower end comes clear of the oil level. The outlet tubes, in contrast, must be narrow enough to prevent air bubbles from rising by relatively greater forces of adhesion. The process works intermittently between an upper and lower point of the oil level.

As Fig. 9 shows, the block diagram of this system presents a closed loop. Philon's oil lamp can thus be described as an elementary example of on-off control. The characteristics of feedback, however, are thoroughly concealed. The device was used again by Heron (see p. 23), by the Banū Mūsā (see p. 44), by Leonardo da Vinci, by Leurechon and his followers (see p. 47), and finally, in modern agriculture, in drinking troughs for animals.

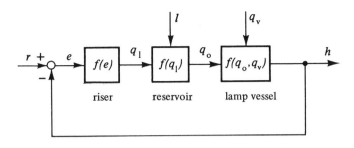

r = desired level: position of lower end of riser
h = actual level
e = air gap
l = level in oil reservoir
q_l = flow rate of air
q_o = discharge of oil from reservoir
qv = oil consumption in lamp

Figure 9 Philon's oil lamp, block diagram.

3. Heron of Alexandria

The works of Heron of Alexandria have been preserved almost completely, notwithstanding their encyclopedic scope, and mostly in their original Greek. But, paradoxically, almost nothing is known about their author's life.[33] Not even his dates are certain. There are several reasons for believing that he lived in the first century A.D. The strongest of these was discovered by Neugebauer in Heron's reference to the lunar eclipse of March 13, A.D. 62 (jul.) which Heron seems to have observed in person.[34]

Heron's works cover most branches of the applied sciences of his time; apart from mathematics, surveying, and optics, his main interest is in mechanical engineering, to which he has devoted books on mechanics, pneumatics, automata, war engines, and water clocks.

Feedback devices are to be found only in the *Pneumatica*. The *Automata* describe complicated apparatus controlled according to a rigid program but not employing any closed loops; the book on water clocks, which might have contained applications of feedback, has been lost except for an insignificant fragment. Heron's *Pneumatica* has been translated into English by Bennet Woodcroft in 1851,[35] but the following discussion is based on Wilhelm Schmidt's definitive Greek-German edition of 1899,[36] which was supplemented by a very detailed volume on the history of the text transmission. Schmidt lists no less than one hundred extant manuscripts of the *Pneumatica*, the oldest and most complete one being the Codex Marcianus No. 516.[37] This manuscript, prepared presumably in the 13th century, was presented in 1468 to the library of St. Mark in Venice by Cardinal Bessarion.

The broad scope of his work prompts the question of how much of it was Heron's original contribution. Style and organization of the *Pneumatica* are occasionally somewhat casual, for which reason Heron has been judged unfavorably. Drachmann maintains, however, that the *Pneumatica* is no finished book but a collection of material that had not been edited for publication. This would explain why in some cases several versions of one device, in stages of development ranging from an arrangement copied elsewhere to Heron's own design, are shown. Heron's devices throughout are mechanically quite refined and practically always fully understood; his presentation is clear. Against slavish imitation he has taken an explicit stand: "Furthermore, one must avoid the predecessors' designs, so that the device will appear as something new."[38]

The devices described in *Pneumatica* I.20, II.30, and II.31 can be regarded as direct derivations of Ktesibios' float valve.[39] In the first case ("Inexhaustible Goblet II," Fig. 10) a float valve is employed to maintain constant liquid levels in two connected vessels. The functions of sensing and control action have not yet been separated. The float serves to measure the actual level and at the same time to control the flow of liquid. The system is therefore equivalent to the arrangement of Ktesibios. The balance beam with the counterweight has only the function of guiding the float properly. Disturbance variables are the level in the supply tank and the rate at which liquid is taken from the regulated vessel.

The apparatus of *Pneumatica* II.31 ("Automatic Wine Dispenser, Controlled by the Rising and Sinking of a Float," Fig. 11) is similar in prin-

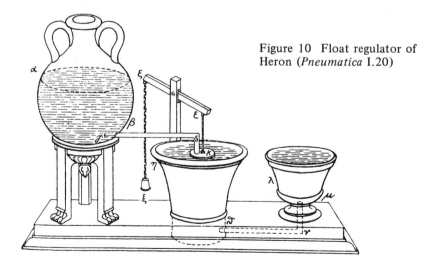

Figure 10 Float regulator of Heron (*Pneumatica* I.20)

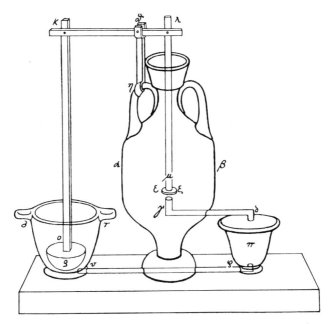

Figure 11 Float regulator of Heron (*Pneumatica* II.31).

ciple to the preceding one but is much improved in technical detail. As before, the problem of maintaining a constant liquid level is solved by a float valve. An innovation, however, is the complete separation of the functions of sensing (float) and that of control action (valve); through this the system has become formally a feedback system in the modern sense. This separation also had practical advantages. The sensitivity and hence the stability of the system can be adjusted by sliding the pivot point ϑ, and the desired level can be set by shortening or lengthening the arm κo. Disturbances are, as before, the static pressure at the valve and the consumption from the controlled vessel. The block diagram (Fig. 12) again shows a closed loop, practically indistinguishable from that of the previous system. Basically, this system is the same as that found in today's boiler drums or flush tanks for toilets.

A variation of the preceding, of similar type, is the "Wine Dispenser Controlled by a Weight" (*Pneumatica* II.30, Fig. 13). The controlled variable here is not the level but the weight of the wine jar. It is measured by a balance where the command variable is easily changed by sliding the weight σ. The comparison between desired and actual weight takes place by a subtraction of torques while the remainder of the device remains unchanged compared to the last.

22 HELLENISTIC TECHNOLOGY

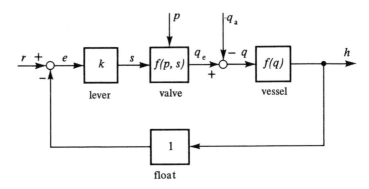

r = desired level (position of float when valve is closed)
h = actual level
s = valve opening
p = water pressure at valve
q_e = flow rate into regulating vessel
q_a = consumption from regulating vessel
$q = q_e - q_a$

Figure 12 Block diagram for Fig. 11 (Heron, *Pneumatica* II.31).

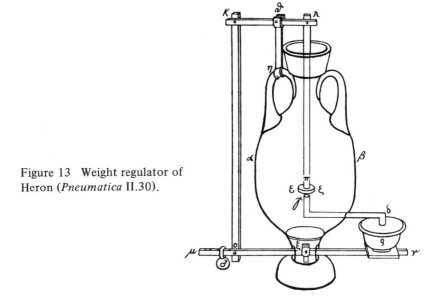

Figure 13 Weight regulator of Heron (*Pneumatica* II.30).

HERON OF ALEXANDRIA   23

A few more of Heron's devices deserve to be mentioned in passing.

In the "Inexhaustible Goblet I" (*Pneumatica* I.19, Fig. 14) we recognize the influence of Philon's oil lamp; Heron himself indicated that he was familiar with the writings of Philon.[40] The device differs from that of Philon (p. 18) only in some details which have been extensively discussed by Drachmann, but which are not relevant to the present discussion.[41]

A favorite device of Heron's is the floating syphon. In *Pneumatica* I.25 (Fig. 15) it appears in an arrangement that shows all the characteristics of a feedback control system. A syphon, connecting two vessels having fluid levels of different heights, floats on the lower of the two, where it is attached to a small overflow vessel. Consequently the liquid will flow from the higher to the lower level until a predetermined level difference is established. The control system regulates this level difference between the two vessels with reference to the lower one of the two. Disturbances would be the addition of liquid to the upper, or the removal of liquid from the lower vessel. The desired level difference is determined by the distance between the overflow ρ and the water line of the float, while the actual upper level corresponds to that in the small overflow vessel. The block diagram (Fig. 16) shows the closed loop. This invention is a feedback device of a significance comparable to that of Philon's oil lamp, but since it is merely a demonstration of a physical principle inapplicable to any concrete task it received no further attention.

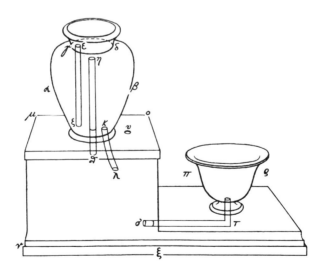

Figure 14 Philon's method of level regulation employed by Heron (*Pneumatica* I.19).

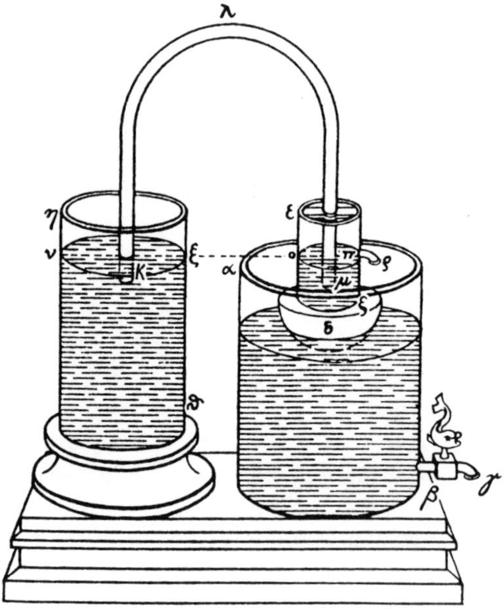

24 HELLENISTIC TECHNOLOGY

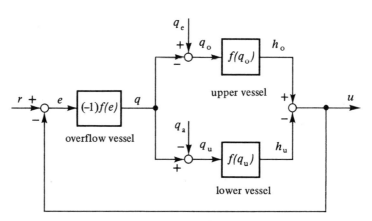

Figure 15
Regulation of a level
difference by Heron
(*Pneumatica* I.25).

r = desired level difference,
distance between overflow and lower level
u = actual level difference
$= h_o - h_u$
e = height of water above overflow

q = rate of overflow
h_o = upper level,
h_u = lower level
q_e = inflow, upper vessel
q_a = discharge, lower vessel

Figure 16 Block diagram of Fig. 15 (Heron, *Pneumatica* I.25).

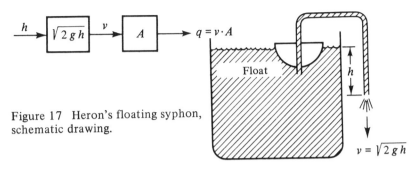

Figure 17 Heron's floating syphon, schematic drawing.

Slight modifications of the preceding apparatus led to the arrangement of *Pneumatica* I.26, whose purpose is still more farfetched. The rate of outflow of wine from one vessel is to have a fixed ratio to the rate of intake of water into another vessel. The closed loop is interrupted, and the formal characteristic of feedback is lost.

A floating syphon in its simplest form is described in *Pneumatica* I.4 (Fig. 17). Since it has the property of maintaining a constant flow rate in spite of a changing upstream water level, one might well ask whether it employs feedback. If so, the controlled variable would be the rate of discharge (proportional to velocity), which is determined by the head h. This variable, however, is not sensed anywhere, not even in disguise. Therefore no closed loop can be identified. The rate of flow is constant, not because of some control action but because of the absence of disturbances. Against obstructions in the tube or unstable floating of the syphon the apparatus would be ineffective.

4. Ancient Feedback Devices—Summary

Heron's feedback devices are more varied and more developed than those of his predecessors. They have similarities with inventions by Ktesibios and Philon, but there is no evidence to indicate any direct dependence. Rather, one might suspect, in late antiquity such devices became known through sources other than the writings of these three authors.

The stimulus for the invention of the float valve, according to Vitruvius' remarks on Ktesibios, seems to have come from the problem of the water clock. Heron's book on water clocks might perhaps have confirmed this presumption. In his *Pneumatica*, Heron had no occasion to discuss water clocks, because it was directly preceded by the volume on water clocks which unfortunately has been lost.

Other areas were ready to adopt the invention, albeit without lasting success because of a lack of suitable practical applications. Both Philon and Heron display a certain delight in the pure principle of this invention, a delight that seems independent of practical use and commercial profit. This attitude differs from the utilitarian outlook of modern technology; but it shows that the Hellenistic mechanicians were able to think in terms of closed causal loops.

One might ask whether devices such as those described here have actually been constructed. There is no archaeological evidence to that effect, but the well-known apparatus found at Antikythera testifies to the remarkably high level of practical precision mechanics in late Antiquity.[42] The lucid style in which Heron has described his apparatus indicates that he must have seen them in physical reality. The craft tradition of water clocks, finally, which survived until the Islamic Middle Ages, would be unthinkable without water clocks that were actually built.

III. Float Valve Regulators in the Tradition of Ancient Water Clocks

In order to follow the subsequent progress of the ancient level control systems we shall pursue two separate tracks: One is the tradition of Heron's *Pneumatica,* the other—to be studied first—will lead us into the history of the water clock.

Its usefulness as a component of the water clock saved the float valve from being considered a mere toy. It represented one of the few methods available in antiquity for achieving the constant rate of water flow that is indispensable for time measurements. Here practical advantages rewarded the labors required to develop a mere idea into a useful appliance. In a practical art such as clockmaking, knowledge is handed down mainly by oral tradition and through the artifacts themselves. Lacking archaeological findings, we are at the mercy of such fragmentary literary sources as have happened to survive.

The literature of late Antiquity and of the early Middle Ages abounds in references to clocks. Most of these are secondary in character, written by authors who had no intention of supplying technical descriptions. Only in a few cases is it possible to reconstruct how the clocks might have worked; water clocks with float valves are not among these. No comprehensive modern study of ancient water clocks is available to date. A great deal of new insight might be gained if the available source material were examined from the philological, technological, as well as from historical point of view.

Besides the numerous secondary references, a few complete books have fortunately survived that deal exclusively with water clocks of ancient tradition. Three of these, the book by Pseudo-Archimedes, the works of al-Jazarī and the book by Ibn al-Sāʿātī (Riḍwān), contain descriptions of water clocks equipped with float valves.

1. The Clock of Gaza

One forerunner of the Islamic monumental clocks is the great magic clock of Gaza described by Procopius in the first half of the 6th century A.D. Unfortunately, his description is not very helpful for our purposes, for it is limited to the exterior of the clock, saying nothing about technical matters.[43]

The clock was located in the center of the city; it was about 6 meters high and 2.7 meters wide, and performed a number of automatic mythological displays every hour. Hercules would fulfill one of his twelve labors, a Gorgon would roll her eyes, and the god Helios would appear. The clock also announced the hours by sound, striking one to six times in two cycles per day. In outward appearance the Gaza clock was similar to the later Arabic clocks, and it is tempting to infer that it also resembled them in its internal workings. Procopius did not mention any water supply line; regulation by float valve is therefore not to be ruled out.

The majority of the clocks mentioned in the literature of the Hellenistic and Roman eras and of the early Islam were located in Palestine and Syria. In these regions the Moslem conquest did not constitute a sharp break, the existing culture living on undisturbed. The traditions of clock construction probably remained unbroken as well. The clock of Gaza, dating from pre-Islamic times, illustrates the continuity from ancient to Islamic clock technology.

2. The Clock of Pseudo-Archimedes

The first of the three great Arabic manuals on horology is entitled the *Work of Archimedes on the Building of Clocks*. It was not written by the great Archimedes of Syracuse, but by some anonymous author to whom we shall refer as Pseudo-Archimedes. It can be dated within the following limits: A *terminus ante quem* is given by the fact that the earliest reference to it is found in the *Fihrist* whose author, Muḥ. b. Isḥāq b. al-Nadīm, died in A.D. 955. A *terminus post quem* is only implied in the Arabic language of the book. The work has survived in three manuscripts (Paris: catalog of G. de Slane No. 2468, p. 437; London: British Museum, catalog of Rieu

No. 1336, p. 619; Oxford: Bodleiana No. 954, p. 95) which form the basis of a German edition prepared in 1915 by Eilhard Wiedemann and Friedrich Hauser.[44] These translators, both physicists by profession, considered themselves unqualified to analyze the book from the philological point of view; the questions of the age of the manuscripts and of the origin of the book itself are still open. Wiedemann and Hauser believed the work to be of Byzantine origin, while Drachmann is inclined to ascribe it to an Islamic inventor. The book's relationship to ancient tradition will be discussed further below.

The work must have had considerable influence: it received favorable comments not only in the *Fihrist* and in the two great horological books by al-Jazarī and Ibn al-Sā'ātī, but also in works by three other Arabic authors. Ibn al-Akfānī (d. 1348), for example, writing on the subject of clocks, states that "the book by Archimedes forms the basis for their understanding."[45]

The description of the complicated clock is so thorough that Wiedemann and Hauser were able to reconstruct it almost completely. The clock was about 4 meters high and had a square base of about 50 centimeters side length (see Fig. 18). It ran for a cycle of twelve hours; the length of the hours was adjustable according to the season. Time was indicated continuously on a dial. Furthermore, every full hour a variety of mechanisms were set into motion. Depending on the hour, a number of marbles would drop from a bird's beak upon a cymbal, a Gorgon's head would roll its eyes, or a snake would frighten a swarm of birds, and so on.

The whole apparatus was driven by a large float (*b* in Fig. 18) on the steadily falling water level in the reservoir *a*. A constant discharge from *a* was maintained by the float valve regulator *cde* which thus constituted the heart of the clock. By keeping the level in the vessel *d* (Arabic: *rub'*) constant, the float valve provided for an even flow through the orifice *g* into the bowl *i*. From there the water would run through the mechanism *jklmnopq*, which periodically activated the birds and the snake, down into the tank *k* at the bottom. The device *fgh* had the function of varying the discharge velocity and hence the duration of the hours by changing the height of the orifice *g*. All the mechanisms in the upper part of the clock were driven by the large float *b*.

The float valve determined the accuracy of time measurement; without it the clock would have been unable to function, for no other provisions were made for lowering the water level at a constant rate.

To demonstrate the correctness of Wiedemann and Hauser's reconstruction, a translation of the original text passages and figures pertinent to the float valve is here quoted (Fig. 19 is a facsimile reproduction of drawings

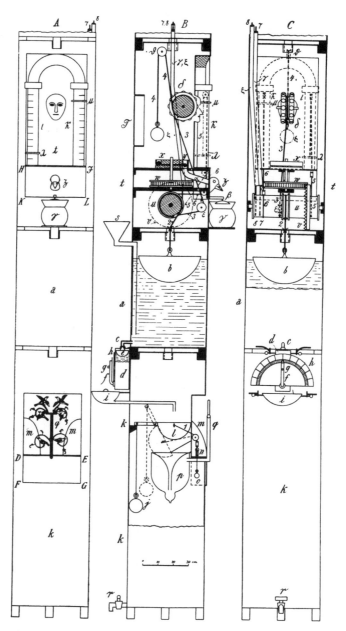

Figure 18 Clock of Pseudo-Archimedes, reconstruction by E. Wiedemann and F. Hauser.

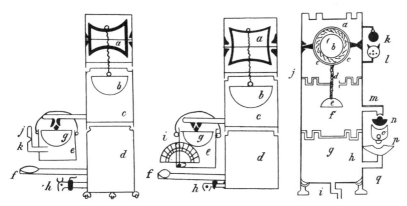

Clock of Pseudo-Archimedes:
Figure 19
According to the Paris manuscript.

Figure 20
According to the
London manuscript.

in the Parisian manuscript, and Fig. 20 shows their counterpart in the London manuscript; the letter designations in the following refer to Figure 18):[46]

Before giving an account of the motions, we shall describe the place where the water is discharged from the tank of the float, and we shall explain in what manner the orifice for the outflow of water is provided. Furthermore we shall show what arrangements are required at this place so that whoever wants to build this clock can proceed accordingly; we shall also describe the shape of the devices, their dimensions, and their positions at their respective places.

One takes a *rub'* [d], similar to the *rub'* of the *kaylajah* [a measure of grain] used for measuring, except that its bottom is convex. On both of its sides it has two buttons. Then one fashions a float [e] that will comfortably fit into this *rub'*. It consists of a semicircle [i.e. a half-sphere] provided with a lid. On its surface is a sort of protrusion that resembles a button.

Then one makes a suitable tube which is inserted into the water tank [a] at a height of about one '*aqd* [the distance between two finger joints] or less from the bottom. It extends into the tank and is half a finger long. It is soldered into the tank. The water of the float tank discharges through this tube. Where the water flows out, the end of the tube is bent downward for the length of one '*aqd*. This bent portion as well as the whole tube are wide enough so that the button on the float [c] can easily fit into it, but it should neither jam inside of it nor should it have too much play.

In function and execution this float level regulator is identical to that of Ktesibios; therefore the comments made earlier regarding its character as a feedback device will hold here as well. To judge from its literary style and the technique of its drawings, the clock book of Pseudo-Archimedes seems to be an Islamic work. The influence of antiquity, however, is evident not only in its claim to have been written by the great Archimedes. The subtitle of the Oxford manuscript reads: "This is what Irūn [Heron] has learned from the books of Philon and Archimedes, the two Greeks, on the hoisting of the weights, the spheres, the waters, the bowls, and so on."[47] Is then Heron considered less Greek than Philon and Archimedes? Near its beginning the manuscript also contains this sentence: "Archimedes has said: My dear Māristūn, I will explain to you, ..."[48] It was Philon, however, not Archimedes, who had dedicated his books to an otherwise unknown Ariston. All this gives the impression that its author had a certain familiarity with the technological literature of antiquity, but that he was not very clear about the relationships between the various authors.

The ancient heritage is even more evident in technical matters. The automatic displays of the clock make use of the same mechanical elements as the apparatus of Philon and Heron. The bird's beak dropping marbles as well as the group of birds with a snake had already been used by the Hellenistic mechanicians. The Gorgon's head also appeared on the clock of Gaza; it stems from Greek mythology. Nor is the clever device for adjusting the length of the hours (Fig. 18, *fgh*) of Islamic origin; according to Vitruvius (9.8.11 ff.) it had been invented by Ktesibios.

It is open to question whether the book as a whole is a translation from the Greek or whether it is an Arabic report on traditional clock technology. Whatever the case, there can be little doubt that the clock of Pseudo-Archimedes as a whole, and the float valve employed in it in particular, represent a legacy from Greek-Roman antiquity.

3. Al-Jazarī

The book of Pseudo-Archimedes had two declared imitators who themselves wrote extensive books on horology, al-Jazarī and Ibn al-Sāʿātī, contemporaries who worked apparently without knowledge of each other. Al-Jazarī appears to have lived after 1181 at the court of the Urtuqids in Amid on the upper Tigris as an instrumentmaker. At his king's behest he wrote an account of the various kinds of apparatus he had built. His book *On the Knowledge of Ingenious Geometrical Mechanisms* was completed in 1206. The first of its six sections deals with the construction of water clocks. A German translation, based upon a manuscript in Oxford (MS. Grav. 27) and two others in Leiden (Nos. 1025 and 1026), was published in 1915 by Eilhard Wiedemann.[49] Wiedemann knew of the existence of seven further manuscripts. This large number of preserved copies testifies to the work's popularity; the unusually high quality of script and illustrations are indicative of a demanding audience. Al-Jazarī confines himself to devices he has built himself. He has stated this expressly many times, but it is also borne out by the character of his presentation. The book's style strongly resembles that of a modern "do-it-yourself" manual, both in the directness of its language and in the realism of its illustrations. The author abounds in practical advice on technological detail, and he cautions against mistakes he has made himself. A typical example for this is his discussion of the device for adjusting the length of the hours on the clock of Pseudo-Archimedes (see p. 29). He systematically tries and rejects various theoretical calibrating scales, finding satisfactory results only through empirical calibration.[50]

The chapter "On Water Clocks" describes ten water clocks that were actually built (Wiedemann added a simple water clock from a later part of al-Jazari's work). The first two of these are expressly modeled after the Pseudo-Archimedan clock. Of the remaining ones, two function by an intermittent process, one consists of a conical vessel of the type of the Egyptian clepsydra, one achieves the constant water level by an overflow device, and the last four are candle clocks. The first two clocks are variations on the same basic principle. Following the Pseudo-Archimedan example they both are regulated by float valves; since this system applies to both clocks, al-Jazarī has described it only in connection with the first one.[51]

The drawing of the complete clock (Fig. 21) does not show the float valve clearly, but al-Jazarī has added to this figure the following legend: "*a* is the main reservoir, *b* the float I, *g* the stopcock whose end is bent downward toward the middle of the *rubʻ*, *d* is the protrusion on float II which rises up into the bent part of the stopcock . . . ."[52]

As in the case of the Pseudo-Archimedan clock, the various time-indicating mechanisms are propelled by float I. A constant outflow from the main tank *a* is assured by the float valve swimming in the vessel ("*rubʻ*") below *d*. From here the water is discharged through the arrangement *ew* for varying the length of the hours, a device which was handed down, as already mentioned, via Pseudo-Archimedes and Vitruvius from Ktesibios. A more detailed description of the float valve is contained in the instructions for making it.[53]

A tube is made of cast copper, half a span in length and wide enough so that an index finger can be inserted into it. A carefully fashioned stopcock is installed in its center. If it is to close, it will close; should one wish to open it, then it will open up.

One end of the tube is bent downward at right angles for the length of about half a finger; the bent end should be wider than the part near the angle. The stopcock is soldered into the lower part of the tank [Fig. 22]. For the bent-down end a conical plug is made of cast bronze; it is fitted

34 ANCIENT WATER CLOCKS

into the bent-down end and ground into it most carefully on the lathe with emery, in the usual manner. This is the procedure for every stopcock and every valve which are to be ground in.

If one places the plug into the bent-down piece without securing it with some object, it will fall down, for its base is wider than its top.

To make the *rub'*, hammer a piece of copper until it takes the shape of the main tank. It is 1½ spans deep and 4 fingers wide. In one side, near the bottom, make a hole, and into this solder a tube of the shape of an index finger [Fig.23]. Then to make the float II [Fig. 24], hammer two pieces of copper until they take the shape of a hollow turnip, and solder them together. Their girth is so dimensioned that this float will easily fit into the *rub'* without touching its sides. The plug is soldered on to the center of one of the circular surfaces of the float. The float then is placed upon a surface of water and carefully observed. If it lists to one side, load the opposite side until it shows no list whatever on the water level.

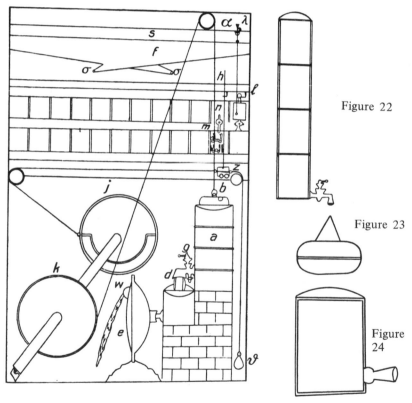

Figure 22

Figure 23

Figure 24

Clock of al-Jazārī.
Figure 21 Over-all arrangement.

Figures 22–24 Details.

This passage limits itself to a step-by-step description of the construction of the device. The actual working of the feedback mechanism is not described anywhere in the whole treatise, except for a vague allusion in a paragraph on the start-up of the clock.[54]

Now water is filled into the main tank a, and the float I [b] is placed upon it. Then one opens the stopcock g. The water flows out from the bent-down end, first into the *rub'*, then into the occlusion at the back of the disc e, then into the tube at the back of the disc. Here the water finds no exit. It rises up in the *rub'*, and so does the float II with the water level until the end of the plug d enters the bent-down end of the stopcock, which it closes; no further water can issue from it. This will happen only after a hole w has been bored into the surface of the disc at a [certain] radius, to serve as the overflow. The water will then flow out through this hole. As water issues from this hole, the float II will go down correspondingly.

These passages, in contrast to Figure 21, show clearly that al-Jazarī employs a float valve. The description, however, is purely procedural and indirect, if not downright awkward. Al-Jazarī does not seem to appreciate the special character of feedback control, in which Philon, Heron, and also the Banū Mūsā had shown great delight.

4. Riḍwān al-Khurasānī, called Ibn al-Sā'atī

The other great treatise on a Pseudo-Archimedan water clock is the *Book on the Construction of the Clock and on its Use* by Ridwān b. Muh. b. 'Alī al-Khurasānī. Its author, commonly referred to by his contemporaries by the title Ibn al-Sā'atī (son of the clockmaker), writes about a clock in the eastern gate (named "Bab Jairūn," i.e., Gate of the Hours) of the main mosque of Damascus. It had been built under the reign of Sultan al-Malik al-'Adil (1146-1173), presumably after a fire in ca. 1168/69, by Ibn al-Sā'ātī's father, after whose death it had fallen into decay. Several attempts to repair it had failed, until Ibn al-Sā'ātī was finally entrusted with it. Although he was neither a clock builder nor a mechanic, but rather a physician and writer, he succeeded in the restoration. For future attendants he wrote an instruction manual for servicing and repairing the clock, which was completed in 1203. Of this work only one manuscript is known to exist, prepared in 1554 in Constantinople and now preserved in Gotha. A German translation was published by Eilhard Wiedemann.[55]

Ibn al-Sā'ātī's style betrays the non-engineer. He is fond of words and avoids illustrations. Even the heavily abridged translation fills 90 quarto pages.

Nothing is known about the book's effect. Ibn al-Sāʿātī himself, however, is mentioned in contemporary Arabic literature: Wiedemann cites a short biography by Ibn Abī Uṣaybiʿah.[56] The clock of Bab Jairūn has been discussed by several Arabic authors, concerning whom Wiedemann has reported in detail.[57] Among them are several eyewitnesses who saw the clock in operation as early as 1184 and as late as 1326; the latter witness did not believe that it worked automatically; he thought a man hidden inside moved the entire mechanism.

Ibn al-Sāʿātī acknowledges frankly and often his dependence upon Pseudo-Archimedes. The opening of his first chapter, for example, contains these phrases:[58]

... and a short summary of the names of individual parts of the clock in question, of which Archimedes has designed several parts: We need not describe its construction [i.e., that of the clock of Archimedes] so completely that we would practically obtain a copy of that scholar's work. We mention only parts he invented, according to the works and accounts of scholars. They are: The main tank, the *rubʿ*, the float II, the cover, the float I, ... The arrangement of the balls, the installation [of the clock], and its general appearance we need not describe as these are mentioned in the man's book.

Evidently Ibn al-Sāʿātī had consulted the book of Pseudo-Archimedes frequently during his work, and he assumed that future clockbuilders would do the same.

The clock's mode of operation can be seen in Figs. 25 and 26. Figure 25 is a reproduction from the manuscript, but Fig. 26 was redrawn by Wiedemann as a composite of several sketches in the original. Clearly visible are the large float I driving the clock, the float valve, and the device for varying the length of the hours.

The float valve shows some improvements. A third vessel (named *kayl*) has been installed between the main tank and the regulating vessel *rubʿ*. It serves to decelerate the flow, first to protect the float valve from being hit by the full force of the on-rushing water and secondly to allow any dirt that may be carried with the water to settle here. The float valve is tied to its valve seat with a string. During operation this string is loosened somewhat, but not so much that the float might become disaligned and miss the valve opening as the level rises. These small improvements appear to be the result of practical experience. In general, the illustrations make it clear that we are dealing with a typically Pseudo-Archimedan float valve.

The text is not as clear as the illustrations. The operation of the control system with its interplay of several elements is never explained. All that is offered is the following instruction for making the float valve and installing it into the clock (Fig. 27):[59]

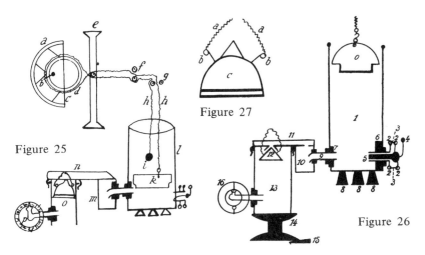

Clock of Ibn al-Sā'ātī
Figure 25 Original drawing.
Figure 26 Reconstruction composed from several original drawings.
Figure 27 Detail: float with valve cone.

Float II is hammered out of a thin sheet of copper; it is hollow, its length is one third of that of the *kayl,* and it barely touches the walls of the *rub',* so that it sinks and rotates without friction. It has the shape of a conical box, being wider at the bottom and tapered toward the top. On its top there is a small cone. The float should be as light as possible; its bottom must be perfectly flat so that it will float on the water in the *rub'.* On top it carries a wide and sturdy plug of cast copper of the same thickness as the opening in the cover II. It is ground into this with emery so to be watertight. Two small buttons are soldered to both sides of this plug, carrying small rings. Tied to these are two silk threads coming from two holes in the cover II. They are fastened tightly at the time when the *rub'* is being attached to the cover and in case no water is desired to flow out. Then, as the threads dry out and the float is buoyed upward, the plug on the float closes the hole in the cover perfectly. When the water is to flow out again, the two silk threads are loosened. If a certain quantity of water issues from the orifice, an equal quantity will flow through the hole in the cover, whereby the plug is withdrawn from the hole in the cover.

The weight of the float must be distributed evenly on all sides, for otherwise the float will list and no longer keep its proper position facing the hole in the cover II. Then more water would flow out than is discharging through the orifice, and the excess would run down the sides of the *rub'.* In hammering out the float, one has to proceed most evenly so that no place will be thicker than another. Pay attention to this!

We may imagine how difficult this device would be to reconstruct without the help of the illustrations, just as in the case of Vitruvius' description of the clock of Ktesibios.

5. Subsequent Fate of the Water Clocks

We have shown that the ancient invention of the float valve remained alive as a component of water clocks until late in the Islamic middle ages. The books of Ibn al-Sāʿātī (1203) and al-Jazarī (1206) are the latest known references to the use of float valves in water clocks. Political and technological events introduced new developments. In the homeland of the ancient water clock, the area between the eastern Mediterranean and upper Mesopotamia, the Mongolian invasion—Baghdad was taken in 1258—brought all creative cultural life to a halt. In Mohammedan Spain, which now became the Islam's intellectual center, float valves do not seem to have been used. New types of water clocks emerged here instead, which did not stem from antiquity and which were clearly superior to the traditional types; with the invention of the mechanical clock these in turn lost their significance.

A detailed account of the state of Spanish horology a few decades prior to the invention of the mechanical clock was put together between 1276 and 1279 at the command of King Alfonso X "the Wise" of Castile.[60,61] Apart from two sundials, a candle clock, and a remarkable compartmented mercury clock whose influence can be traced into the 17th century,[62] it describes a water clock in which a constant flow rate of water is achieved by the principle of Mariotte's bottle. Similar to Heron's floating syphon, this is accomplished not by the action of feedback but simply by the absence of disturbances.

Arrangements such as these were rendered obsolete in the 14th century by the invention of the mechanical clock. From that time on the water clock carried on a marginal existence. Some inventors, among them Leonardo da Vinci, searched in vain for some principle that would enable the water clock to compete with mechanical clocks.[63] In 1552 Taqī al-Dīn, a Syrian living in Ottoman Constantinople, described in his book on clocks the state of horological affairs quite accurately:[64]

I have classified the clocks into three groups:

1. The sand clocks [i.e., hour glasses] are generally known and at everybody's disposal. They are of no great use, for only rarely do they give the same timing in both directions. . . .
2. The water clocks, of which there are many kinds. . . . They too are of no great use; indeed they cause more trouble than benefit.

3. These clocks are the ones with which we will deal. They are the revolving clocks . . . .

With the demise of the water clock the float valve lost its chance to be carried into the modern era by this path. Between 1206 (al-Jazarī) and the mid-18th century no references to float valves are known.

IV. Float Valves in the Tradition of Heron's *Pneumatica*

The career of Heron's *Pneumatica* can be traced through the centuries with remarkable continuity.[65] From the late middle ages on the work attracted wide attention; more than one hundred manuscripts of it are preserved, all prepared after the 12th century.[66] Less is known, however, about its influence in the preceding period. No direct references to the *Pneumatica* are known in the literature of pre-Islamic times, even though such authors as Pappus, Patricius, Proklus, and Eutokius have frequently cited other works by Heron. There is only indirect evidence of a critical edition in the 6th century. One group of manuscripts of the *Pneumatica* is clearly distinguished from the others by a number of corrections and deletions that were made with a certain liberty. On the basis of their Greek style, Wilhelm Schmidt estimated them to date from the 6th century, terming their originator as Pseudo-Heron.[67] The latter must not be confused with the anonymous compiler of the 10th century known as Heron of Byzantium or Heron the Younger, who has written on military technology, surveying, and sundials, but not on pneumatics.[68] In the early middle ages Heron's *Pneumatica* found a sympathetic audience only in the Islamic world. This is demonstrated by two works written in a truly kindred spirit: al-Jazarī's book on ingenious mechanisms, which has already been described in the context of water clocks, and above all the *Kitāb al-Ḥiyal* of the Banū Mūsā of the 9th century.

1. The Banū Mūsā

The three brothers Banū Mūsā (the Sons of Mūsā) lived in the 9th century at the court of the Abbasid caliphs in Baghdad. Owing to their prominence in science they became high state officials, and thus their significance is twofold: Their scientific work (a list prepared by F. Hauser contains 21 titles) dealt mainly with mathematical and physical subjects. For example, they are the authors of a treatise on geometry, known as *liber trium fratrum*, that was translated into Latin in the 12th century by Gherardo di Cremona in Toledo. Perhaps better known yet is their contribution to technology, the *Kitāb al-Ḥiyal (The Book on Ingenious Mechanisms)*, written in the vein of the *Pneumatica* of Philon and Heron, the subject of this chapter.[69]

Equal in importance to their scientific accomplishments was one particular administrative activity of theirs. A prerequisite for the rise of Arabic science was access to the sources of antiquity. In the first half of the 9th century, great efforts were made to translate Greek literature into the Arabic. In this the Banū Mūsā had a leading part. They employed their political power as well as their personal wealth to purchase books and to hire translators. They sent out agents; it is known, moreover, that Muhammed, the eldest, personally journeyed to Byzantium in this pursuit. Two of the translators employed by them have become famous, Ḥunayn ibn Isḥāq (809-877) and Thābit ibn Qurra (826?-900);[70] also Qusṭā ibn Lūqā, who later translated Heron's *Mechanica*, did this under their influence. It is not known precisely when Philon's and Heron's books on pneumatics became available in Arabic. Most of Heron's work had probably already been translated by Hunayn ibn Isḥāq.[71] It is exclusively to such Arabic translations that we owe our knowledge of Heron's *Mechanica* and Philon's *Pneumatica*. Both Philon and Heron became widely known; several Arabic encyclopedists have devoted articles to them.[72]

In the *Kitāb al-Ḥiyal* the brothers' interest in ancient Greek literature combines with their own work in the field of mechanics. Philon and Heron are never mentioned by name, but the book's contents are clear evidence of their influence. The *Kitāb al-Ḥiyal* is a collection of the descriptions of one hundred pieces of pneumatic and hydraulic apparatus. Conceptually they are based on a relatively small number of ancient inventions such as the syphon, the float valve, Philon's oil lamp, or the water wheel. The Islamic contribution consists of technological refinement—Arab metal workers were highly accomplished—and of devising new applications. Most of the devices were drinking vessels with some special feature, occasionally quite far-fetched, designed to awe the unsuspecting user. Only a few are useful in the modern sense, but they are all expressions of pure delight in the clever exploitation of physical phenomena.

The *Kitāb al-Ḥiyal* must have been widely known in the Islamic world. Hauser has listed a number of references to it in Arabic literature,[73] of which the following remark by the encyclopedist al-Anṣārī (d. 1311) is a typical example:[74]

42   FLOAT VALVES

The science of the pneumatic machines . . . The best known work on this science is the famous book on the ingenious mechanisms by the Banū Mūsā. There is a compendium on this by Philon and an extensive work by Badi' [al-Zaman] al-Jazarī.

The translators of the *Kitāb al-Ḥiyal*, E. Wiedemann and F. Hauser, knew of two manuscripts, one located at the Vatican (No. 317), the other partly in Gotha (Pertsch cat. No. 1349), partly in Berlin (Mq. 739 Ahlwardt No. 5562), and partly in Leiden (Cat. vol. 3, No. 1019, 169 Gol.).

Among the hundred devices in the *Kitāb al-Hiyal* we find four applications of the principle of Philon's oil lamp (Nos. 76, 84, 95, 97) and eight float valves modeled after those of Heron's (Nos. 75, 77-83).

a. Float Valve Regulators. The arrangement No. 75 (Fig. 28) is characteristic for this group.[75] Water runs from the river F through the pipe R with the valve H into the trough T. A float S in the trough is connected with the valve by the linkage S-s-P so that the valve will be opened with a falling float and vice versa, thus maintaining a constant water level in the trough. The arrangement is proposed for use as a drinking vessel for farm animals.

The device seems to be directly inspired by the float valve in Heron's *Pneumatica* (Fig. 11). The only difference, apart from the changed application, is the design of the valve. Heron had improvised a valve by holding a flat plate against the end of the pipe (Figs. 9, 11, 13). This was adequate only when used under low water pressure. The constant force required to keep the valve closed had to be supplied by the float.

Throttling valves installed directly in the pipe and requiring no constant force to keep them closed were not employed in Heron's designs. In connection with float regulators they appear first in the book of the Banū Mūsā. This constitutes a genuine step forward. It enables the apparatus to work under higher water pressures, eliminating the risk of valve leaks due to an insufficient closing force or a misalignment of the valve plug. The brothers' book shows several other progressive valve designs. In principle, however, this first level regulator is equivalent to that of Heron's *Pneumatica* (Fig. 11), and both systems have identical block diagrams (Fig. 12).

The apparatus No. 78 (Fig. 29) describes a different application of the same conception.[76] Instead of water from a river it regulates the liquid contained in a tank.

The devices Nos. 77, 79 and 81 belong into the same group.[77] They are equipped for an additional task unrelated to feedback control: the valve will close not only when the vessel is full but also when it is empty. The arrangement No. 77 (Fig. 30) accomplished this by means of an opposed-

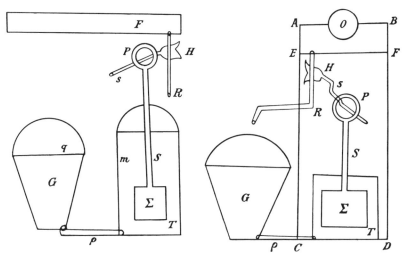

Figure 28   Device no. 75.

Figure 29   Device no. 78.

Figure 30   Device no. 77.

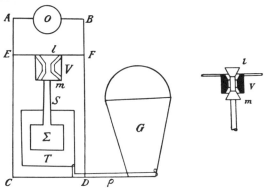

Float regulators of the Banū Mūsā.

plug double-seat valve of modern appearance, in which the two valve plugs face each other from different sides of the valve seat. The arrangements Nos. 79 and 81 employ different valves of which, however, no details are given. The inventors suggested the following applications. The invention could be employed in drinking troughs for small animals, which would discourage the fast-drinking larger animals by promptly running dry when used by them. On the other hand it might serve to intoxicate unsuspecting drinking companions. If it was the objective of a drinking contest to drain the wine vessel, the initiated would drink quickly, because the vessel would not refill once it was empty. The uninitiated guest, in contrast, drinking slowly, would not finish until he had emptied the vessel as well as the hidden reservoir.

The arrangements Nos. 80 and 82 employ float valves of the same type, combined with auxiliary effects of no interest in this context.[78]

b. On-Off Control. Of basic significance, however, are the systems Nos. 83 and 84 (Figs. 31-34).[79] A constant liquid level is achieved here in the following manner. When the level reaches an upper limit, the supply is stopped and remains stopped even after the level begins to sink. Only when it has reached a lower limit will the supply be opened again to refill the vessel. The control action is not steady: it never reaches a point of equilibrium; instead the level fluctuates between an upper and a lower limit.

The unsteady mode of operation does not disqualify the system as a feedback device. It still has the purpose of maintaining a given variable at a prescribed value; this value, however, is defined by two limits rather than by a single point. The system forms a closed loop; the block diagram of No. 83 does not differ from those of Nos. 75 and 78. The function of sensing, performed in No. 83 by a float, is clearly distinguished from that of corrective action, the valve. Systems of this class are widely used in modern technology; they are commonly referred to as "on-off" control systems.

The Banū Mūsā achieved such control in two ways: In device No. 83 there is a modification of the simple arrangement of No. 75 whereby the linkage between the float and the valve has been changed to create a dead zone (Figs. 31 and 32). In No. 84 the same thing is accomplished without moving parts by combining the principle of Philon's oil lamp with some cleverly arranged syphons (Figs. 33 and 34). The vessel *ABEF* is a reservoir, filled through the aperture *o*. The level to be regulated is that in vessel *G*.

c. The Principle of Philon's Oil Lamp. In addition to the example just described, the Banū Mūsā have applied Philon's method of level control to three other cases.

The arrangements No. 95 (Fig. 35) and 97 are obvious imitations of Philon's oil lamp (Figs. 7 and 8), although with genuine improvements.[80] An ingenious combination of syphons has been added to the original system by which the lamp can easily be refilled even while it burns. In the apparatus No. 97, the oil lamp No. 95 is combined with a lamp described by Heron in *Pneumatica* I.34, where the wick is advanced automatically by a float on the level of the oil reservoir. In No. 76 finally the syphon regulator is combined with some sort of hydraulic toggle switch;[81] this somewhat obscure design holds nothing new from the point of view of

HERON'S PNEUMATICA   45

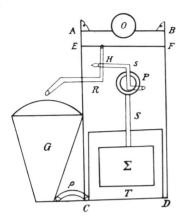

Figure 31   Device no. 83.

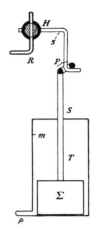

Figure 32   Detail of device no. 83 (reconstruction by E. Wiedemann and F. Hauser).

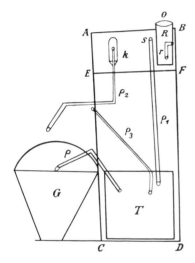

Figure 33   Device no. 84.

Figure 34   Device no. 84, schematic drawing by Wiedemann and Hauser.

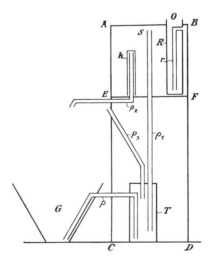

On-off regulators of the Banū Mūsā.

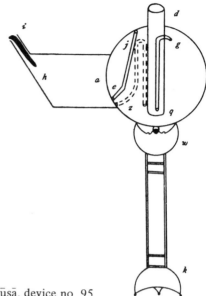

Figure 35 Oil lamp of the Banū Mūsā, device no. 95.

feedback control. Incidentally, an oil lamp similar to that of Philon is also described by al-Bīrūnī.[82]

2. Heron's *Pneumatica* after the Banū Mūsā

While our knowledge about the early career of Heron's *Penumatica*, as outlined earlier (p. 40), is very fragmentary, we are well informed about the work's transmission from the Christian middle ages on. According to Henricus Aristippus, the *Pneumatica* had been available in Sicily at the time of King William I "the Bad," i.e., in the middle of the 12th century. The work is also said to have been translated by William of Moerbeke.[83] The oldest manuscript extant, the *codex marcianus no. 516*, presented by Cardinal Bessarion to the library of St. Mark in 1468, is believed to date from the 13th century.[84] In the following two centuries numerous additional manuscript copies were made in the original Greek, and a few also in Latin translation. References to Heron are found in discussions of the vacuum problem in the *Summa philosophia* by the Pseudo-Grosseteste (13th century) and by Giovanni Fontana (15th century).[85] The interest in Heron's *Pneumatica* increased by the end of the 15th century. Regiomontanus had projected its translation from Greek into Latin, and in 1501 Giorgio Valla, an Italian humanist, included excerpts of it in Latin in his work *De expetendis et fugiendis rebus,* by which Leonardo da Vinci presumably was introduced to the *Pneumatica*. Heron's influence is also evident in Girolamo Cardano's *De Subtilitate* (1550). On the whole, during the Middle Ages and Renaissance, the work was both well known and accessible; yet it had only limited influence.

A drastic change took place in 1575, when Federigo Commandino's Latin translation of Heron's *Pneumatica* appeared in print. A vast increase in the book's audience was due to two factors: the printed book was incomparably cheaper than the traditional hand-written copy; and while Greek was read only by a small elite, Latin was the common language of educated Europeans. In 1583 Commandino's translation was reprinted. Furthermore, Italian editions were published by Bernardo Davanzati in 1582, by Battista Aleotti in 1589, by Alessandro Giorgi in 1592 and 1595. The result was a veritable Heron renaissance. This manifested itself in the sudden appearance of a large number of books on machinery, all by authors who were more or less overtly inspired by Heron: among them are Guidolbaldo del Monte (1577), Strada (ca. 1580), Ramelli (1588), Giambattista della Porta (1589), Lorini (1597), Zonca (1607), de Caus (1615), Veranzio (1616), Fludd (1617), Leurechon (1624), Branca (1629), Schwenter (1636), Kircher (ca. 1640-70), and Schott (1657). Only now, in the baroque era, were Heron's inventions finally accepted in Europe. They have contributed significantly to the achievements of 17th-century technology.

What part do the ancient level regulators play in all this literature? A crude version of Philon's oil lamp appears at the beginning of Cardano's *De subtilitate libri XXI* (1550) and, adopted from that source, in Giambattista della Porta's *Pneumaticorum libri tres* (1601); it is discussed extensively—still in this form—in Robert Hooke's *Lampas* of 1677. An authentic version of Philon's lamp reappears first in Leurechon's *Récréations mathématiques* (1624), from where it is promptly copied by Kaspar Ens, Daniel Schwenter, and Caspar Schott. It has not been established how Cardano and Leurechon had found access to Philon's book, but there is evidence that Latin translations from the Arabic of Philon's *Pneumatica* had been available in Renaissance Europe.[86] On the evolution of feedback control this device had no appreciable influence. Not only was it limited in use; it was unsuited as a demonstration of the principle, since the action of feedback is completely concealed in the operation of the oil lamp.

Far more significant is the fate of the float valve regulator. Anyone who searches for applications of this invention will be sorely disappointed. The float valve as applied to level control is totally absent from the technological literature of Europe between the 12th and the 18th century. This is difficult to believe, but to verify this observation, practically all relevant literature in this time span was carefully reviewed, from the writings of Herrad of Landsperg, Villard de Honnecourt, Guido da Vigevano, Konrad Kyeser, Francesco di Giorgio Martini, Leonardo da Vinci, Vannoccio Biringuccio, Georgius Agricola, Robertus Valturius, Jacques Besson, and the above-mentioned authors of the Heron renaissance, to the works of Jakob Leupold and the great encyclopedists of the 18th century.

The float valve first emerges again in England in the middle of the 18th century, not in literature but in industrial practice (see p. 76) where it had been reinvented, apparently independently of Heron's *Pneumatica.*

Why was the float valve so utterly ignored in Europe up to the 18th century? The period between the late Middle Ages and the baroque era was certainly not lacking in mechanical ingenuity or technological skill. Moreover, after Commandino's translation of the *Pneumatica* had appeared in print, a large number of readers were directly confronted with the invention. It is inconceivable that it was accidentally overlooked or misunderstood. One is forced to conclude that the float valve level regulator was consciously rejected by the engineers of the Renaissance and Baroque.

V. Automatic Control in Ancient China

The size and quality of the contribution of Ancient China to technology suggest the question whether any feedback mechanisms were invented here. A survey of Joseph Needham's great work[87] has not produced enough evidence for an affirmative answer, but a few ancient Chinese inventions deserve to be examined here even if they fail to make use of feedback.

1. The South-Pointing Chariot

The south-pointing chariot was hailed by Needham as the "first homoeostatic machine in human history"; it is therefore fitting to discuss it first.[88] The southpointer is a two-wheeled chariot carrying an upright human statue. By means of a special mechanism (perhaps a differential gear) this figure continually points southward, regardless of the orientation of the chariot. It thus enables the traveler to keep a straight course without depending on external landmarks. Such a chariot has been mentioned by several ancient Chinese historians; both its date (12th century A.D. or earlier) and its reconstruction are still being debated. The question of whether the chariot is a feedback device in our sense—Needham uses the adjectives *cybernetic* and *homoeostatic* interchangeably (homoeostasis is a biological feedback system)—can be decided without considering its internal construction or practical success.

The purpose of the device (like that of a compass) is to indicate the actual orientation of the vehicle. The indication is read by the driver and compared with the desired direction. It is the driver who then initiates the necessary course corrections. What results is a directional control system where the functions of comparison and control action are fulfilled by a human, in other words a system with manual feedback (see p. 7). Needham is fully aware of this. He calls the south-pointing chariot "... the first homoeostatic machine in human history, involving full negative feedback. Of course, the driver has to be included in the loop."[89] The second sentence cancels out the first. Any human action can be interpreted as a case of feedback control, in which the human performs the functions of sensing and corrective action. There is no difference in principle between crossing a desert with the help of the south-pointing device, and simply walking up to a tree. The south-pointing chariot thus is no feedback device by our definition.

2. The Drinking-Straw Regulator

Next we turn to a feedback loop that again includes a human element, but with a difference. Here man does not have the controlling functions of sensing and corrective action; instead, he plays the role of the disturbance. The purpose of the device is to limit arbitrary human actions: it regulates the wine consumption of the participants in a drinking bout. The report on the drinking straw regulator of the Ch'i Tung natives in southern and southwestern China is given in the travel journal *Ling Wai Tai Ta* by Chou Ch'u-Fei (died 1178):[90]

They drink wine through a bamboo tube two feet or more in length, and inside it has a 'movable stopper' (kuan li). This is like a small fish made of silver. Guest and host share the same tube. If the fish-float closes the hole the wine will not come up. So if one sucks either too slowly or too rapidly, the holes will be [automatically] closed, and one cannot drink.

The device, which may be visualized as in Fig. 36, was designed to enforce a uniform rate of drinking, i.e., to maintain a constant rate of flow. The block diagram (Fig. 37) shows the principal difference between it and the south-pointing chariot. The desired flow is determined by the weight of the float. If the upward draft is too strong, the hydrodynamic lift overcomes the weight of the float; the float will rise, stopping the flow of wine. A maximum flow rate of wine, however, will be obtained by a drinker who does not draw as hard as he can, but who instead optimizes the suction by intuitively finding the derivative $dq/dp = 0$. At this suction pressure a steady flow rate will establish itself. The drinking-straw regulator therefore again does not represent genuine feedback control, for it achieves equilibrium only when used by a man who, by a process of optimization, forms a second closed loop.

FLOAT VALVE REGULATORS   51

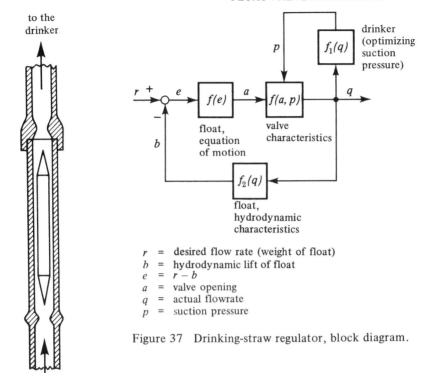

Figure 37   Drinking-straw regulator, block diagram.

Figure 36   Drinking-straw regulator.

Historically the drinking-straw regulator is unique. The Ch'i Tung natives were quite isolated from the influences of foreign cultures because of their geographical situation; the invention must be their own. In Europe, control devices of this type were patented, perhaps for the first time, by Achille Elie Joseph Soulas in 1841 (Brit. Pat. No. 8894).

3. Float Valve Regulators

Did the highly developed Chinese water clocks employ float level regulators? In principle Needham does not believe so.[91] Two passages exist that seem to refer to such regulators, but on closer examination they prove to deal with something else.

The accurate running of the famous monumental clock of Su Sung (ca. A.D. 1090) was based upon the continuous flow of water from a constant-level tank upon the water wheel. The water tank is described in Su Sung's text as follows (translated by Needham and Wang):[92]

The upper reservoir and the constant level tank: The latter has a water-level arrow [i.e., marker]. The manual wheel forces the water to flow into the upper flume and so to the upper reservoir. As the water enters the reservoir at a nonuniform rate, the water-level tank is made to adjust [i.e., the pressure head]. After escaping, the water flows to the scoops of the great wheel. Since the water remains uniformly [in motion] throughout the day and night, time can thus be correctly kept.

The original illustration accompanying this does not add anything toward an understanding of this passage.[93] Needham asks whether the "water-level arrow" is to be identified as a float valve, as in the clock of Ktesibios. The text offers no evidence for this. More likely the pointer served purely as a level indicator for the clock's attendants.

The other case in point is Wang P'u's water clock (ca. A.D. 1135), described in a 13th century edition of Yang Chia's book *Liu Ching T'u* (first published about 1155).[94] Here several compensating vessels are combined with an overflow vessel in an arrangement that makes a float level regulator dispensible in any case.

An indicating rod swimming on the lowest level does have the characteristics of a float valve (according to the text; the book's illustration does not show this), but its purpose is not to maintain a constant level; it only stops the water supply to the clock at the end of the cycle when the measuring receiver is filled.

Part Two
Modern Times after 1600

VI. Temperature Regulators

1. The Thermostatic Furnaces of Cornelis Drebbel

The first feedback system to be invented in modern Europe and independently of ancient models is the temperature regulator of Cornelis Drebbel (1572-1633) of Alkmaar, Holland.[95] A mechanic and chemist, he spent the main part of his life in the service of Kings James I and Charles I of England. Between 1610 and 1612 he served at the court of Emperor Rudolf II in Prague, together with his famous contemporary Johannes Kepler. He became first known for a perpetual-motion device: actually it was a thermoscope which would maintain a continuous although slow motion due to the temperature fluctuations during the course of the day. Then, as a lens grinder and instrument maker he was among the first who have built telescopes and microscopes. As a chemist he found pyrotechnic materials, discovered a most successful process of scarlet dyeing, and allegedly knew how to produce oxygen. Finally, he is said to have built a submarine and to have carried out successful dives in it. In the following, however, our attention is directed toward his furnaces equipped with automatic temperature control.

Drebbel, who generally wrote little, left no records concerning the thermostatic furnaces. Enough contemporary secondary sources are known, however, to convince us that the invention is Drebbel's own, and to tell us how it had looked and worked.

The only reference to the thermostatic furnace during Drebbel's lifetime is a diary entry by N.C. Fabri de Peiresc (1580-1634) in the year 1624. He wrote about Drebbel:[96]

Et s'adonna à alchimie, où il dict qu'il a trouvé des choses miraculeuses, et des inventions de fourneaux admirables entres aultres un pour conserver le feu en pareil degré de chaleur selon que l'on le desire plus ou moins ardant.

He devoted himself to alchemy, in which he claims to have uncovered miraculous things, and inventions of wonderful furnaces, among them one that will maintain the fire at any degree of heat desired, whether hotter or colder.

De Peiresc, a learned gentleman from Aix-en-Provence who had once studied in Padua under Galilei and who is noted mainly for his correspondences with prominent scholars (Bayle called him "le procurateur général de la littérature"),[97] was not personally acquainted with Drebbel; he owed his information to Drebbel's son-in-law Abraham Kuffler who had visited him in Paris.

Soon after Drebbel's death, Hildebrand Prusen and Howard Strachy, as executors of his estate, took out an English patent (No. 75, A.D. 1634) on Drebbel's furnaces on behalf of his descendants.[98]

C[er]tayne Stoves and Furnaces ..., the Heate may thereby be Increased, Moderated, or Abated to any Pr[op]orcion or degree that shalbee held most Fitt or Requisite for any the Uses aforesaide, with much lesse Charge, shorter Tyme, lesse Attendaunce ..., then heeretofore hath bin donne by any other.

The formulation of the patent is awkward and perhaps intentionally vague, but it refers, at least between the lines, to the thermostatic properties of the furnaces. A generation later, the invention was still remembered by men like Boyle, Hooke, and Wren. Robert Boyle (1627-1691) wrote in 1660:[99]

Yet it is certain, that Drebble, that great, singular, learned mechanick ... found out a furnace which he could govern to any degree of heat; but whether these have died with him, or how far the meditations of others have wrought upon them, I shall humbly refer to a more leisurable enquiry.

He had no clear conception of the temperature regulator, nor did he feel any great urge to find out more about it. Robert Hooke (1635-1703) too must have known of it, as Tierie has shown.[100] But he has not expressed himself upon it; instead, he later proposed a competing method (see p. 66) that achieves constant temperature without resorting to feedback.

The young Christopher Wren (1632-1723), a professor of astronomy who had no idea of his future career in architecture, and who entertained himself by inventing self-writing thermometers, precipitation and wind meters, was more openminded toward Drebbel's thermostats. The French traveler De Monconys, after a visit with him, recorded in his diary on June 11, 1663:[101]

Il me dit aussi sa pensée pour faire un fourneau comme celuy de M. Keffer [= Johan Sibertus Kuffler], scavoir qu'il y ait devant le Registre un Vase, qui soit moitié dans le fourneau et moitié dehors, et qui soit plein d'argent-vif; lequel se haussant lors que l'air de la cornue qui est sur les cendres, le presse, il bouche le registre; car la muraille du fourneau est comme un diaphragme qui divise le vaisseau du vif-argent en deux, comme cette figure en fera souvenir.

He also told me his idea for making a furnace like that of Mr. Keffer [= Johan Sibertus Kuffler], namely, that in front of the damper hole there was a vessel, half inside the furnace and half outside, which was full of quicksilver; this will rise when the air in the retort, which is resting on the coals, presses against it, and it will close the damper. For the furnace wall is like a diaphragm which divides the quicksilver vessel in two, as this drawing will recall.

Drebbel's invention was also mentioned in the transactions of the Royal Society (October session, 1662):[102]

Sir Robert Moray offered to the consideration of the society a way to compare the effects of heat and cold in rarefaction and condensation of air, with that of force or weight. Upon which Dr. Goddard suggested Drebbel's method of governing a furnace by a thermometer of quicksilver.

All these quotations clearly attribute the invention to Drebbel, but they offer no technical details.

The role of an executor of Drebbel's intellectual inheritance had been adopted by his second son-in-law, the physician Johan Sibertus Kuffler (1595-1677) of Cologne, who devoted his life to the effort of extracting commercial profit from Drebbel's inventions. Apparently he had built several specimens of the thermostatic furnace, according to the testimony of such eyewitnesses as J. Moriaen and Sorbière in Holland[103] and de Monconys in England.[104] He is also the source, at least indirectly, of detailed technical information about the temperature regulators, which has been transmitted in two ways.

I. In 1666 Augustus Kuffler wrote a manuscript with the title "A very Good Collection of Approved Receipts of Chymical operations..."[105] Augustus Kuffler, son of Johan Sibertus and hence grandson of Drebbel, was a physician like his father. The work is a collection of all recipes, processes and inventions that the Kufflers thought worthy of preservation, and it was probably based on Drebbel's own laboratory records which according to de Peiresc had passed into Johan Sibertus' possession.[106] A detailed description of the furnaces, accompanied by two illustrations (Figs. 38 and 39) is found in the first of the manuscript's five parts:[107]

58 TEMPERATURE REGULATORS

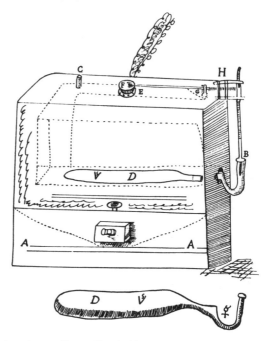

Figure 38 Drebbel's incubator (from Cambridge manuscript).

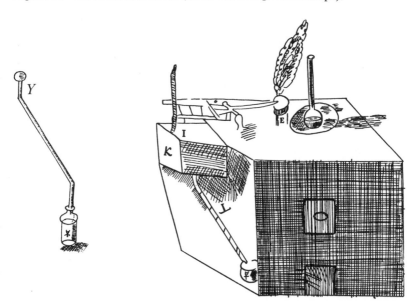

Figure 39 Drebbel's Athanor (from Cambridge manuscript).

DREBBEL'S FURNACES 59

The Description of two Furnaces ruling themselves and Keeping at any Degree of Heate the one for Hatching of Chickens.

The First Figure ... is for the curious Chymist the 2nd Figure ... is for Hatchin of Chickens and must Bee tended with greate circumspection and difficultie this Furnace must Bee without A grate having 2 or 3 holes running from the Place where the Δ (fire) is to the Edges which Blow the Δ as A over the fire Lyes An Iron Plate with A hole in the middle as B: thro which the Heate comes over which is placed A Double foure Square Tinnen, Leaden or Copper Box in which the Eggs are Laid in Towe and with in the Double sides Bottome and Top the ∇ (water) is put with which it must Bee filled thro a small Pipe Comeing out of the Topp of the Fournace as C and still as the ∇ water wasts itt must Bee filled Againe thro the same withe Bottome of the water Box Between the Double of itt Lay the glass D: which is filled with spiritus vini to the Neck and in the Necke $\ogon$, to fill this Retorte you must first Put the $\ogon$ (mercury) in then the spiritus vini then Turne itt upp side downe, holding to the mouth and if $\ogon$ will come into the neck Let the Δ come round the square water Box and it must Come out att a round small hole in the middle of the Topp of the Furnace as E upon which you must have a spoone to shut as F which spoone must have A long handle playing upon A Cross pinn at G and at H It hath A Screw By which meanes It may Be fitted Backward or forward, now there must Be another Pinn with A Screw att the end of which is Put A Little glass Pipe and fitted into the Neck of ye retorte as 1: Soe that when the Δ groweth hotter the Ordinary the spiritus vini expands ittselfe pressing upon the $\ogon$ and the $\ogon$ the Pinn I. and so closeth the hole E and dampe the Δ till It comes to A iust heate, the first Figure hath noe ∇ Box But the glass retort D: is Laid and Built in the wall of the Furnace Leaving itt Bare to the heate A fingers Breadth all a Long nor must itt have spiritus vini in it but Emptie and ye neck with $\ogon$ as ye other which is with the Place K. The Barromeeter, to Know what Degree of heat the Furnace is at, Being A small Bolts Head with A Long Neck Bent Crooked neer the Bowle put in to the side of the Furnace and the other Bent Crooked and put into A violl of $\ogon$ and as ye heate increases the Aire will rarifye and the $\ogon$ will goe downe when you will Sett itt att any degree of Heate as to make itt Hotter Screw upp the Pinn I and soe the spoone Keepes the Opener the Pipe of the Bolts Head guides you what Degree itt is att.

The language of this text, apart from the obvious archaisms, is so awkward and faulty that one is tempted to hypothesize it might represent the own words of Drebbel who had learned English only when in his thirties. The contents of the description, however, are quite clear.

II. The other source of technical details is the travel journal of Balthasar de Monconys. De Monconys (1611-1665), a nobleman from Lyon with good connections at the court in Paris, traveled a good deal in Europe and in the Near East. It was his custom to call on every notable scholar in hopes of learning the latest concerning his researches. In 1665/66 he published the results of his travels in a *Journal des voyages* with the characteristic subtitle "... *où les Sçavants trouveront un nombre infini de nouvautez, en Machines de Mathématique, Experiences Physiques, Raisonnements de la belle Philosophie, curiositez de Chimie, et conversations des Illustres de ce Siecle."* (... where learned men will find an infinite number of new things, as mathematical machines, physical experiments, philosophical discussions, chemical curiosities, and conversations with the great men of this century.) The book was successful, it had new editions in 1677 and 1695, and in 1697 it was even translated into German. In the report on a trip to England Drebbel's name appears frequently; and the thermostatic furnace is mentioned not only in the conversation with Christopher Wren (see p. 56). On June 2, 1663, de Monconys, accompanied by Henry Oldenburg, the Secretary of the Royal Society, paid a visit to Johan Sibertus Kuffler where he was shown the thermostatic furnace. He was sufficiently impressed to include in his *Journal* a lengthy description of it accompanied by an illustration. In style it is more polished than the Kuffler manuscript, but less comprehensive in content. Nevertheless, both are in complete agreement: The illustrations, too, are strikingly similar (compare Fig. 39 with Fig. 40). De Monconys' description reads as follows:[108]

Il a un autre fourneau de Philosophe que i'ay veu, lequel estant plus eschauffé que l'Artiste ne desire, sans que personne y touche, il fait baister une palette dessus un registre, qui en estant ainsi fermé, la chaleur diminue, iusques à ce qu'il soit au degré qu'il desire; et si le feu estoit trop foible, cette mesme palette se leue, et le registre fournissant de l'air au feu, il reprend la vigueur et le degré necessaire. Cest instrument est en dehors à un costé du fourneau, et à deux ou trois pouces plus bas, il y a un tuyau de verre ioint contre la muraille du fourneau, incliné de quelque 25. degrés, gros comme une plume; au bas *duquel* il y a du vif-argent: le haut n'est que de l'air, lequel s'eschaufant trop, fait descendre le vif-argent, et ainsi l'on voit par des marques qu'il y a au tuyau, la quantité de feu. Et quand le feu est trop lent, et que l'air se condense, le vif-argent monte et marque par sa hauteur le degré qu'il y a de froideur. En voicy à peu pres la figure.

Explication du Fourneau [See Fig. 40]
A   Registre.
B   Platine de fer qui le bouche.
BC Branche de fer qui porte la platine.
EE Cheualet de fer sur lequel la verge *BC* balance en equilibre.
F   Ouuroir ou Sablier à mettre les Vaisseaux.
C   Extremité du bout de la verge *BC,* qui est un anneau escroüé dedans.
D   Fer tourné à vis, qui entre dans l'anneau C, qui fait tousjour l'equilibre avec *B,* en quelque facon qu'on le mette; mais qui doit disposer l'effet de l'artifice, selon le plus ou moins de chaleur qu'on desire; car s'il est fort enfoncé dans
I   Trou où entre ce fer virollé dans la capacité du fourneau c'inclination de la verge *BC,* en estant moindre, la moindre chaleur le fera mettre en

DREBBEL'S FURNACES   61

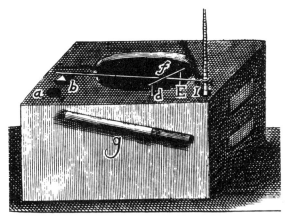

Figure 40 Drebbel's Athanor (according to de Monconys).

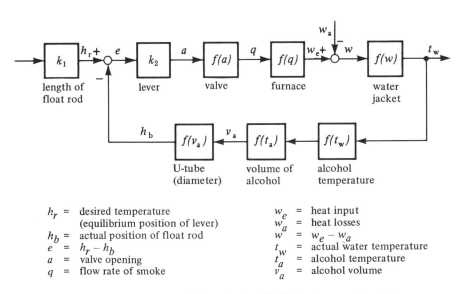

h_r = desired temperature
       (equilibrium position of lever)
h_b = actual position of float rod
e = $h_r - h_b$
a = valve opening
q = flow rate of smoke

w_e = heat input
w_a = heat losses
w = $w_e - w_a$
t_w = actual water temperature
t_a = alcohol temperature
v_a = alcohol volume

Figure 41 Temperature regulation in Drebbel's incubator, block diagram.

equilibre, partant la platine couurira d'abord le registre: au lieu que s'il est peu enfoncé, l'inclination de la verge *BC,* estant plus grande, il faudra plus de mouuemement pour la remettre paralelle, situation necessaire pour boucher le registre.

G Est le tuyau de verre avec du vif-argent en bas et de l'air en haut, du quel les deux bouts entrent dans la capacité du fourneau.

There is another laboratory furnace which I saw, which, when hotter than the operator wants, without anyone touching it lowers a lid over an opening. When this is thus closed, the heat is reduced until the desired temperature is reached; and if the fire be too weak, this same lid rises, supplying air to the fire, so that it is revived and reaches the desired temperature. This device is outside, to one side of the furnace and two or three inches lower. There is a glass tube attached to the wall of the furnace at an angle of about 25 degrees, about as thick as a pen, at the bottom of which there is some quicksilver, above which there is only air which, when overheating, will push down the quicksilver. Thus one can tell the amount of fire by looking at the markings on the tube. And if the fire be too slow, and the air condenses, the quicksilver will rise and by its height indicate the degree of coolness. And here is an approximate picture of the furnace:

Explanation of the furnace [see Fig. 40]:
A Aperture
B Iron lid that covers it
BC Iron rod that supports the lid
EE Iron fulcrum on which the iron rod is balanced
F Opening, to hold vessels
C End of the rod *BC*, which is in the form of a ring threaded inside
D Screw which passes through the ring *C*, which is always in equilibrium with *B* no matter how it is adjusted, but which must regulate the mechanism depending on whether more or less heat is desired. It is firmly embedded in
I Hole into which fits the screw that goes into the furnace. If the angle of the rod *BC* is small, the least bit of heat will put it in equilibrium, provided that first of all the lid covers the opening. If the screw is slightly tightened, the angle of the rod *BC* being greater, more motion is needed to make it parallel, which is necessary in order to cover the opening.
G The glass tube with quicksilver at the bottom and air at the top, the two ends of which go into the interior of the furnace.

The Kuffler manuscript deals first in great detail with the incubator (Fig. 38); more briefly, it mentions a furnace for the use of chemists in general, a so-called "Athanor" (Fig. 39). This second furnace was seen by De Monconys, and described in his journal (Fig. 40). The temperature regulation works as follows:

In the incubator, the fire $A-A$ in the lower third is separated from the oven chamber by an iron plate with an opening in the middle. The oven chamber contains an incubator box, drawn in dashed lines, which has hollow walls filled with water through the feed pipe *C*. The smoke rises past the walls of the incubator box before escaping through the opening *E*. Inserted in the water-filled bottom of the incubator is the temperature feeler *D* (also shown in Fig. 38, bottom), a peculiarly shaped glass tube which protrudes through the right side of the stove. Its cylindrical part at the left is filled with alcohol; the U-shaped open part on the right contains mercury. The mercury level in the open branch of the U tube rises and falls (due to the thermal expansion of the alcohol) with the temperature of the water. The vertical rod *I* floating on the mercury is connected above with the lever *H* which is pivoted at *G* and which carries the damper *F* at its other end. The system works thus: if the water in the incubator becomes hotter than desired, the flue opening *E* will be closed by the damper *F*. The fire, receiving less air, will therefore produce less heat. It is noteworthy that both the rod *I* and the lever *H* are threaded at the ends. This provides adjustability both for the desired temperature (the command signal) and for the sensitivity of the feedback loop.

The chemical furnace (Figs. 39 and 40) serves merely to heat a retort; it therefore has no complicated inserts. Its temperature regulator is similar as that of the incubator; however, its temperature feeler, hidden in the box *K*, is filled with air instead of with alcohol, so as to be able to operate at higher temperatures. The furnace is further equipped with the thermometer *Y* (drawn separately in Fig. 39, left) to indicate the temperature inside the box *K*. This thermometer consists of an S-shaped, air-filled glass tube whose open end is immersed into the mercury. The mercury level at the bottom of the tube will fall with rising temperatures.

Figure 41 represents a block diagram of the temperature regulator on the incubator. As already noted, both the desired temperature and the proportional sensitivity can be adjusted easily. Disturbances, such as fluctuations in fuel supply and heat consumption, may occur. Particularly remarkable is the construction of the temperature feeler by which the feedback loop is closed. It is arranged as a chain of many links: glass tube—alcohol—mercury—floating rod—lever with flue damper. The chain is far too long to have been found fortuitously. The inventor seems to have grasped the fundamentals of feedback quite generally before he applied them in practice.

According to all available evidence Drebbel must be regarded as the inventor of temperature regulation and hence as the inventor of the first feedback mechanism of the West. A genuine invention, the formulation of an unprecedented technically realizable idea, is a historical discontinuity. Can Drebbel, who is comparatively unknown today, be credited with such an intellectual feat? It is difficult to appraise him since he has left few written works. Unlike some less significant contemporaries he did not content himself with publishing his inventions in books; instead, he transformed them into concrete—if perishable—reality. We know of his other inventions, as of his furnaces, only through the reports of third parties, and we regard such secondary testimony intuitively as less convincing. Drebbel's efforts covered a wide area, but they all are original as well as practicable—even the alleged *perpetuum mobile*. Drebbel combined a rich imagination, characteristic of the Baroque, with uncommon sobriety and independence of judgment. By means of empirical intuition he was able to advance far beyond the academic science of his time in a number of instances. Indirect evidence of Drebbel's intellectual caliber is the great number of prominent men who valued his acquaintance, and the unrestrained praise that he drew from Robert Boyle, Constantin Huygens (Christiaan's father), and the painter Peter Paul Rubens.

If it is conceded that Drebbel was intellectually equipped for an important invention, then it remains to examine the external conditions.

Drebbel lived in a period of extreme scientific activity. Among his contemporaries we find Bacon, Galilei, Kepler, Descartes. Moreover, Drebbel was always in close touch with the spirit of his time. This not only holds true for his life in the two capitals London and Prague. In his native town of Alkmaar he was a fellow citizen of a number of notable scientists, such as the brothers Adriaen (1571-1635) and Jakob (1591-1623) Metius, the former of whom is known as the teacher of Descartes, and Jan Leeghwater (1575-1650) who drained the Netherlands after they had been flooded during the Wars of Liberation. In London, Drebbel is known to have been acquainted with Francis Bacon, Henry Briggs, Robert Fludd, Constantin Huygens, Peter Paul Rubens, and probably with Salomon de Caus.[109] It may also be assumed that during his years at the Imperial Court in Prague (1610-1612) he had met Kepler and Jost Bürgi.

A scientific prerequisite for Drebbel's temperature regulator was the knowledge that the quantity of heat generated in a fire depends on the quantity of air available. Furnaces with adjustable air dampers have been mentioned first in England about 1500; after that date they are occasionally but not frequently encountered on stoves.[110] What actually gave rise to the invention? The motive may have come from alchemy. Apparently Drebbel believed that transmutation, i.e., the conversion of a base metal into gold, might be feasible if the initial mixture was heated at a moderate but constant temperature for a very long time. The thermostatic furnace was intended to satisfy this requirement.[111]

Heron's *Pneumatica*, for the first time available in understandable Latin and inexpensive print, now reached a wider audience. It gave an important

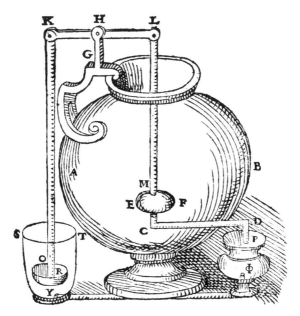

Figure 42   Heron's float regulator (*Pneumatica* II:31), according to Commandino.

impetus to the inventive activity of the engineers of the time, and it must have contributed to the increased interest in pneumatics and hydraulics. Drebbel, who lived in two of Europe's leading capitals and who was acquainted with so many prominent scientists and inventors, must have been familiar with the *Pneumatica*. It was probably here that he found the principle of the thermoscope, which he employed in several of his inventions including the temperature regulator. Heron's float valves may have suggested to him the possibility of regulating a temperature instead of a water level. Indeed a certain similarity can be observed between the regulator on Drebbel's incubator and Commandino's illustration of the float valve. In both cases a float is connected to a valve by means of an arrangement of pivoted levers (cf. Figs. 38 and 42).

2. Further Temperature Regulators
in the 17th century: Schwenter, Hooke, Becher

According to reports, Drebbel's temperature regulator has actually been built and, given proper maintainance, should have worked. But it was not an instrument for industrial use that could be marketed in larger numbers.

66   TEMPERATURE REGULATORS

The temperature feeler, consisting of an open glass tube filled with alcohol and mercury, must have been extremely delicate. Apart from the aforementioned statements by men from the circle of the Royal Society we have no information on the subsequent fate of the invention. The Kuffler manuscript remained unpublished, and the idea of temperature regulation lay dormant in England for over a hundred years.

Nevertheless, there must have been a certain demand for thermostatic furnaces during the 17th century. Especially chemists—lacking the Bunsen burner—must have felt an acute need for a source of constant heat. Beside Drebbel's invention several other methods were proposed.

In 1636 Schwenter revealed "a beautiful secret for a chemist, how to keep heat always at the same degree" (*ein schön Secret für einen Chymicum, die Hitz immer in einerlei Grad zu erhalten*).[112] He solved this problem without feedback. Aqua fortis (nitric acid) trickled slowly and steadily from a capillary tube upon a piece of iron where it generated heat of reaction at a suitable rate.

A similarly naive procedure was proposed in 1677 by Robert Hooke who, as a friend of Boyle and Wren and an acquaintance of the Kufflers, must have been familiar with Drebbel's regulator. In his little book *Lampas* he describes several oil lamps made to burn evenly by various arrangements of counterpoises which maintain a constant distance between oil level and wick.[113] Feedback was not used. Higher temperatures, if desired, were achieved simply by employing the necessary number of lamps simultaneously.

The invention of a genuine temperature regulator is claimed by the versatile Johann Joachim Becher (1635-1682).[114] In 1680, at the end of a restless life as a physician, chemist, and economist, he came to England where he wrote two books containing this claim. His little volume *Närrische Weissheit und weise Narrheit (Foolish Wisdom and Wise Folly)* of 1682, after discussing Drebbel's thermoscope (*not* the thermostatic furnace) contains this statement:[115]

> ... und ich habe 2 Usus von der Thermoskopia erfunden. Eines, dass damit in einem Chymischen Ofen gantz gleiche Wärme regieren kan, dann das Thermoscopium selbsten ziehet das Ventil, wordurch die Hitze in den Ofen gehet, nach verlangter Proportion auff und zu, lässet also mehr oder weniger Hitze hinein, welcher Gestalt man sehr stet Feuer geben kann.
>
> ... and I have invented two uses for the thermoscope. One, by which in a chemical furnace very constant heat can be maintained. The thermoscope itself opens or closes the valve, through which the heat passes into the oven, in the desired proportion, thus admitting more or less heat, in which fashion one can maintain a very constant fire.

This describes the principle of closed-loop temperature control clearly enough. A similar claim had already been made by Becher in his *Minera arenaria perpetua* (1680), the third supplement to his *Physica subterranea*,[116] in which he claimed to have seen such a furnace in operation, and to have published the invention as early as 1660. However, he supplies neither technical details nor illustrations. Earlier references to temperature regulation in Becher's printed works have not been found. It is difficult to decide whether Becher's claim to have invented the temperature regulator independently should be accepted. He has no witnesses for it (indeed his ethics have occasionally been questioned), and he has made the claim only at a time when he had easy access to information about the Kuffler's furnaces in England.

3. Réaumur and the Prince de Conti

While the unpublished invention was soon forgotten in England, de Monconys' widely read travel journal presented the temperature regulator to a large audience on the Continent. Here the idea continued to develop until it became general knowledge of engineers in the early 19th century. During this entire career the temperature regulator was closely connected with the practical application of the incubator.

The next witness of this development was the versatile physicist René-Antoine Ferchault de Réaumur (1683-1757).[117] In the last decade of his life he experimented extensively—motivated partly by a wish to improve agricultural methods and partly by embryological and genetic interests—with the artificial hatching of chickens,[118] on which he reported in several publications between 1747 and 1757.[119] His incubators were deliberately simple, for he sought methods that could be used by ordinary farmers. A distinguished visitor, the Prince de Conti, on seeing the furnaces, suggested an improvement that is indeed equivalent to Drebbel's temperature regulator. Réaumur described this proposal accurately, in courtesy to the Prince, or in delight over the invention.[120] Réaumur's report is doubly significant. It not only indicates reliably that such regulators had been unknown up to then—as long-term director of the *Académie Royale des Sciences,* Réaumur was in touch with most of the prominent scientists of his time—it also proved to be instrumental in the survival of the invention.

Technically, the Prince's regulator does not go beyond that of Drebbel. He had suggested to insert into the interior of the stove some device that would open an air vent as soon as a certain temperature was reached so that warm air could escape and cold air enter in order to bring the temperature down. Réaumur summarized the method elegantly as "employer ces degrés de chaleur contr'eux-mêmes, pour les faire agir pour se détruire . . ." (making use of these degrees of heat against themselves, so as to cause them to destroy themselves).[121]

Next, Réaumur discussed several designs of temperature feelers that could be employed in the regulator, starting with two arrangements proposed by Jacques de Romas (1713-1776).[122] In the first, the thermal expansion of a simple iron bar is amplified by appropriate levers to such a degree that it could adjust a damper. In the other, a bottle partly filled with liquid is suspended horizontally in an unstable equilibrium. When the liquid expands at rising temperatures, the equilibrium of the bottle is disturbed, causing a motion which is then used for the control action.

Réaumur foresaw better immediate prospects for two other arrangements, both based on the thermal expansion of liquids or gases in a bottle in the neck of which a plug moved up or down as a measure of temperature. The Prince had a model of the latter type built, but Réaumur does not supply any details of it. Such temperature feelers were bound to be troublesome; if the plug did not seal well, they would be inaccurate; if the plug was too tight, they might fail to move the damper. Such devices might work under laboratory conditions, but they would not be sufficiently reliable for commercial use. Réaumur wrote sceptically: ". . . des machines de cette espèce ne seront jamais, selon toute apparence, employés que par des curieux" (. . . it appears that machines of this type will never be employed except by the curious).[123] This was quite true as far as the device in question was concerned; but he overlooked the possibility of a more suitable temperature feeler.

The "Prince de Conti," whom Réaumur has credited with the invention, can be identified from the context as Louis-François de Bourbon (1717-1776), a most remarkable man.[124] At the time when he occupied himself with Réaumur's incubators, in his early thirties, he had already concluded a brilliant career as a military commander culminating in an adventurous campaign over the Alps into Savoy and Piedmont (1744-1745). During the decade following the Prince was a confidential secretary to Louis XV and a foe of Mme. de Pompadour, with a great deal of influence on French politics. Later he lived away from the court in the company of such men as Rousseau, Beaumarchais, and Diderot. He has been praised as one of the most enlightened spirits of the century. Perhaps he had reinvented the temperature regulator independently; it seems more likely, however, that he had earlier read de Monconys' travel journal, and that from there he had remembered—if only subconsciously—Drebbel's invention.

4. William Henry's "Sentinel Register"

Another temperature regulator was invented in colonial America at the eve of the Revolution. Its inventor was William Henry (1729-1786) of Lancaster, Pennsylvania, a well-known gunsmith who made a hobby of experimenting in his own laboratory with various uses of steam power. On a business trip to Britain he reportedly met James Watt; he built a steamboat, which however was not successful. During the Revolutionary War he served in various public offices. He was a close friend of the instrument maker and astronomer David Rittenhouse, and a supporter of the painter Benjamin West and of Robert Fulton, the builder of the first successful steamship.[125]

When the newly founded "American Philosophical Society" of Philadelphia in 1771 published its first volume of transactions, William Henry contributed a description of a temperature regulator he named "Sentinel Register."[126] The apparatus is basically similar to those of Drebbel and Réaumur, but it differs in the design of the temperature feeler. The original illustration (Fig. 43) shows the system applied to a furnace in which a flue damper A is controlled by the temperature feeler D. A reconstruction of the latter, based on Henry's detailed descriptive text is given in Fig. 44. The air in the copper vessel C expands with rising

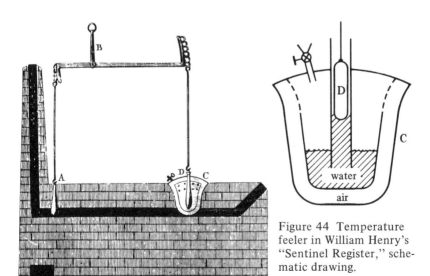

Figure 44 Temperature feeler in William Henry's "Sentinel Register," schematic drawing.

Figure 43 William Henry's "Sentinel Register."

temperature, and by exerting pressure on the water level it forces the water to rise in the vertical pipe. The float D, through appropriate linkages and levers, confers this motion upon the damper A in the flue. The stopcock allows the water to be replenished in the vessel C. Henry states that he has operated the apparatus successfully for well over a year, recommending it for use in chemical furnaces, in the manufacture of steel and porcelain, in hothouses and incubators, and for the temperature control in hospitals.

The Sentinel Register has a great deal of similarity with Drebbel's temperature regulator. The temperature feeler in both consists of a float moving on an intermediate fluid according to the thermal expansion of a gas, and acting upon a flue damper. All circumstances seem to indicate that Henry's Sentinel Register is part of a tradition deriving from Drebbel's invention. Henry may have been inspired through de Monconys' or Réaumur's books, but he may also have heard directly of Drebbel's work on this visit to England.

5. Bonnemain

The leap from an improvised laboratory device to a rationally engineered practical appliance was performed in the temperature regulator of the French engineer Bonnemain (ca.1743–after 1828). While Drebbel's invention remained practically unknown, surviving only through scattered remarks in books on totally different subjects, Bonnemain's apparatus was described in the leading French, English, and German journals, reaching practically everyone interested in scientific technology.

Little is known about Bonnemain's life. His inventions include, in addition to the temperature regulator, a system of hot-water heating based on natural circulation.[127] Even his first name is unknown, and his life dates are only implied in the information that in 1828, at the age of 85, he received the silver medal of the "Société d'encouragement pour l'industrie nationale."[128] He spent most of his life in Paris, and he referred to himself as "ingénieur physicien." In 1777 he announced to the "Académie royale des Sciences" an observation concerning the hatching of young chickens.[129] In 1783 he received a patent for his temperature regulator.[130] Between ca. 1778 and 1794 he must have operated a rather progressive chicken farm in the outskirts of Paris, where by means of thermostatic incubators he raised large numbers of chickens which were supplied to the royal court and to the market in Paris. The enterprise was given up during the Revolution when feed ran short and the clientele dwindled away. In 1816 Bonnemain published a pamphlet "Observation sur l'art de faire éclore et d'élever la volaille sans le secours des poules," praising the economic advantages of his incubators and quite explicitly soliciting orders for their construction. Technical details were kept secret. This appeal presumably found little response, for a few years later he accepted from the Ministry of Interior a grant of money ("encouragement pécunaire") on condition that he would turn over drawings and descriptions of his apparatus to the Société d'encouragement for publication in its *Bulletin.* At the Exposition of 1823 he received applause for some of his mechanisms.[131] In 1828 he was awarded the aforementioned silver medal in recognition of his contribution to the method of heating by natural hot-water circulation. It would be worthwhile to investigate the many gaps remaining in Bonnemain's biography further.

We may assume that Bonnemain came upon the idea of temperature regulation via the detour of chicken farming. His interest in raising chickens dates back to the 1760s.[132] Studying the literature on the subject he must have become familiar with the work of Réaumur who was the leading authority on artificial chicken hatching. Here he presumably read the report on the regulator of the Prince de Conti, from which he took the principle and the initiative for his own design. Bonnemain submitted a complete specimen of his *régulateur de feu* to the Académie royale des Sciences which in 1782 gave it full approval, predicting for it a successful future.[133] The patent specifications of 1783 were not published; all technical details remained unknown until 1824, when the *Société d'encouragement* published drawings of the temperature regulator.[134] Here the device was installed on the furnace of a hot-water heating plant. Later Bonnemain's artificial incubator as a whole appeared in the *Dictionnaire technologique,* including the same furnace and regulator.[135]

The following description is based on the article of 1824, bypassing matters related to heating, and concentrating on the feedback mechanism (Figs. 45 and 46). Figure 45 shows the over-all arrangement: the temperature feeler (dotted lines) is immersed in the water jacket of the furnace. The control affecting the process of combustion is activated by moving the air register at the bottom of the furnace. The regulator attached to the top connects the temperature feeler with the register. The element crucial for the success of the apparatus, the temperature feeler, is based on a principle already implied in the proposal of Jacques de Romas, as quoted by Réaumur. Bonnemain achieves a truly effective temperature feeler by arranging two rods made of metals of maximum difference in their coefficients of thermal expansion (i.e., lead or zinc against iron) to act against each other (Fig. 46a).[136]

The iron rod xx which is firmly mounted at the top $a'h'$ is concentrically surrounded by a tube of lead or zinc. This tube is closed at the bottom by a copper plug y, into which the lower end of the iron rod is screwed from

72 TEMPERATURE REGULATORS

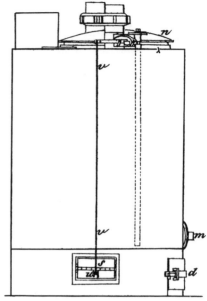

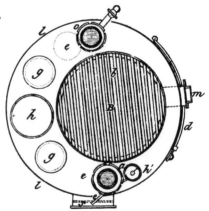

Figure 45
Bonnemain's temperature regulator, overall view.

inside. A copper ring z is soldered upon the upper rim of the tube. This ring will move upward with rising temperatures, because the thermal expansion of zinc or lead is more than twice that of iron. This motion is received by two levers $b'd'e'$ which are arranged in series, amplifying it by a factor of about 100 so that a temperature change of 30°C will cause the outer end of the lever e' to move downward by ca. 4 centimeter. This end of the lever is connected with the air register of the furnace by a rod v

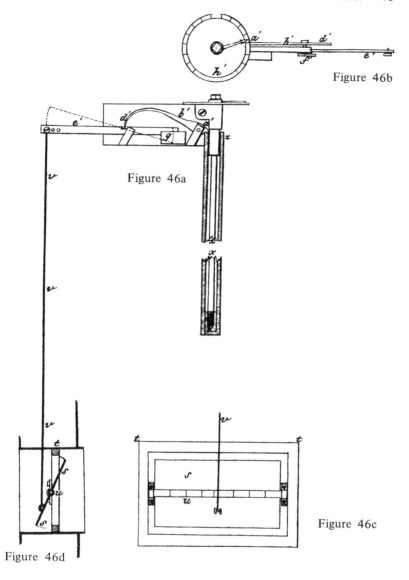

Figure 46 Bonnemain's temperature regulator, details.

(Fig. 46c and d). When the water temperature has risen above the desired value, the register closes, the furnace produces less heat, and the water temperature will fall.

The apparatus is equipped with a device for setting the desired temperature:[137] the iron rod xx at its upper end a' is mounted in a screw device (Fig. 46a). If rotated, the whole rod will move up or down, shifting the zero position of the regulator. The angular position of the iron rod is indicated by the pointer i' on the scale h' (Fig. 46b). A possibility to adjust the sensitivity of the regulator is indicated by the two free holes at the end of the lever e' where the rod v is attached.

The various sources on Bonnemain assert repeatedly that the regulator proved valuable in practical use. On examining the drawings of the apparatus one is inclined to accept this estimate. Its conception is faultless. The temperature feeler is sufficiently sensitive and at the same time strong enough to overcome the resistance of friction, inertia, and gravity. Above all, the apparatus is mechanically stable. Unlike the thermoscope, it will not lose a given calibration or adjustment except through external damage. Furthermore, the device is carefully designed in all details. For example, the two levers b' and e' amplifying the temperature reading (Fig. 46a) are combined with a counterpoise g in such a manner that the mechanism always moves under a positive force pointed in the same direction; this eliminates the danger of backlash caused by the unavoidable play in the bearings at c and f. A definitive evaluation of Bonnemain's temperature regulator would of course require actual tests.

Reports about Bonnemain's invention spread unusually far and quickly, as the following, doubtless incomplete, survey will show. As already mentioned, Bonnemain received a patent in 1783, but it is not known how it was published. In 1786, *Lichtenberg's Magazin* published a short notice about Bonnemain's regulator,[138] without giving details or source. This was reprinted in 1821 in *Busch's Handbuch der Erfindungen.*[139] A similar short notice in the Krünitz encyclopedia, 1812 added that this regulator was "already quite forgotten" (schon ganz in Vergessenheit geraten).[140] After the *Bulletin de la Société d'encouragement* published a full description of the temperature regulator in 1824, hardly half a year passed before a literal translation of it appeared in *Dingler's Polytechnisches Journal.*[141] The article on Bonnemain's incubator in the *Dictionnaire technologique* of ca. 1827 was taken over promptly and fully by Dingler's journal[142] and by *Gill Technological Repository.*[143] Also Andrew Ure's *Dictionary of Arts, Manufactures and Mines* of 1839, which became widely known in many editions and translations (a German translation by Karmarsch and Heeren appeared in Prague, 1844), included an extensive treatment of Bonnemain's apparatus, now under the term "thermostat."[144] An incubator by Bonnemain of 1815 was listed in a 1910 catalog of the *Conservatoire National des Arts et Métiers* in Paris,[145] but in 1966 it could not be found.[146]

If Bonnemain's invention was not immediately adopted by industry, it was the fault neither of technical shortcomings of the device, nor of insufficient publicity. Nor can it be reasonably maintained that there were no suitable applications. During the course of the century more and more new temperature regulators were invented. For example, in 1839 Andrew Ure presented, in his *Dictionary,* beside Bonnemain's construction several of his own (patented in 1831) which in turn may have inspired other inventors.[147] The subsequent history of temperature regulation has been dealt with by Ramsey.[148]

VII. Float Valve Regulators

1. The Reappearance of Float Valve Regulators in the 18th Century

The float valve of antiquity had not only survived in the tradition of Arabic clock building until as late as the 13th century, but after 1575 it had also become generally accessible in the West through the various translations of Heron's *Pneumatica*. It is therefore rather odd that, as pointed out before (p. 48), the float valve was consistently avoided by the inventors of the Christian West from the Middle Ages to the Baroque, and that it reappeared only in the 18th century, seemingly independent of ancient tradition. We first encounter it in England in the level regulation of house water reservoirs and of steam boilers. In the domestic application float valves seem to have been introduced in England around 1740; it was a useful invention, since city houses were supplied with running water only periodically, about three times a week for two to three hours.[149] William Salmon in his book *The Country Builder's Estimator* (3rd edition, 1746) listed the price of "Ball-Cocks, the Ball six inches in Diameter, the Cocks one inch, at 12s. each";[150] they were marketed at fixed prices in standard sizes.

Such float valves were a prerequisite for the invention of the water closet during the last quarter of the century. Before 1800 ten British patents were issued dealing with W.C.'s, the most important being those of A. Cumming (No. 1105 of 1775) and J. Brahmah (No. 1177 of 1778). None of all these disclosed how the level of the water tank above the toilet was regulated. Apparently this was considered generally known and nonproblematic.

2. James Brindley

The earliest known application of a float valve regulator in steam boiler construction is described in a patent of James Brindley (1716-1772), who had started as a millwright and ultimately became famous as an engineer of bridges, canals, and aqueducts.[151] Between about 1756 and 1758 he constructed a steam engine in Fenton Vivian, Staffordshire, in the course of which he invented a number of improvements.[152] To protect these, Brindley in 1758 took out British Patent No. 730, which carried the title: "A Fire Engine for Drawing Water out of Mines, or for Draining of Lands, or for Supplying of Cityes, Townes, or Gardens with Water, or which may be applicable to many other great and usefull Purposes in a better and more effectual Manner than any Engine or Machine that hath hitherto been made or used for the like Purpose."[153]

The detailed description of the invention (no drawings are included) deals with the construction of an improved steam boiler. The item of interest for us is found in the following sentence: "At the top of the said boyler is to be fixed a pipe called the feeding pipe, which, by means of buoys in the boyler that rise and fall with a clack, supplys the boiler with water without attendance; but if the part be neglected by persons who attend on common fire engines, there boyler is often damaged, and sometimes destroyed."[154]

The feed valve (*clack*) is controlled by the float (*buoy*), and the whole arrangement works automatically: it amounts to nothing less than a float valve regulator of the type of those of Heron and the Banū Mūsā. Since Brindley had personally supervised the construction of a steam engine, one may assume that he was not describing an armchair invention but one that he had tested in actual practice.

3. I. I. Polzunov

A float valve of considerable individuality was independently invented in 1765 in Russia, under similar cimcumstances. Ivan Ivanovich Polzunov (1728-1766), an engineer at a coal mine at Barnaul, Altai (Siberia), from 1763 to 1766 constructed a steam engine to drive fans for blast furnaces.[155] The engine worked according to Newcomen's principle, but in its details it was designed quite independently. It had two cylinders; instead of a walking beam it employed a certain unconventional transmission system, and it could be completely disassembled into easily transportable components. Polzunov died a few days before the engine was put into operation. It worked successfully for some time, then it seems to have fallen into disrepair; it was finally forgotten until 1882, when A. N. Voyeykov accidently discovered Polzunov's papers in Barnaul.

Polzunov equipped his steam boiler with a float level regulator; unfortunately, the original illustrations[156] are so indistinct that the device can be understood only with the aid of a schematic drawing from a modern Russian textbook (Fig. 47).[157] The arrangement (guidance of float, valve design) is sufficiently different from the ancient and also from contemporary English designs that it may have been invented by Polzunov independently. In its principle of action, i.e., in the block diagram, Polzunov's device does not differ from the known older systems.

4. Sutton Thomas Wood

Chronologically the next reference to a float level regulator for steam boilers can be found in British Patent No. 1447 of 1784. It was granted to the brewer Sutton Thomas Wood of Oxford (he also received three other patents in 1784, 1785, and 1792), and bears the title: "Certain new Improvements on the Steam Engine, and also the Method of adapting and connecting the Coppers or Boilers employed in the Operation and for the Purposes of Brewing to any Engine or Engines that are worked by the Powers of Steam and Air, so as to render the Steam produced from the Boiling Works and the Steam produced during the Operation of Brewing capable of Working those Engines commonly known by the name of Fire or Steam Engine."[158]

The extensive (23-page) patent specification, accompanied by excellent drawings, describes nine improvements, each consisting of numerous suggestions concerning the design and application of steam engines and boilers. The first group, dealing with improvements on boiler armatures, includes the float valve (Fig. 48). The accompanying text is rather wordy but adds nothing to the drawing; therefore it need not be repeated here. It is perhaps remarkable that it already refers to the float valve, near the end of the description, as a *regulator*.[159] The word *regulator* later became the common term for closed-loop control devices of all kinds, a predecessor of our term *feedback control*. In technical detail Wood's regulator shows a striking similarity with that in Heron's *Pneumatica* II.31 (Fig. 11). We notice, incidentally, that the same invention is protected here for the second time (after Brindley) by a British patent. Among the older British patents it is not uncommon that a given invention is covered more than once.

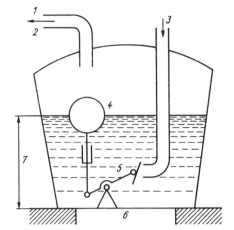

1 steam discharge
2 disturbance
3 feed water
4 float
5 valve
6 furnace
7 actual level

Figure 47 Boiler level regulator of I. I. Polzunov.

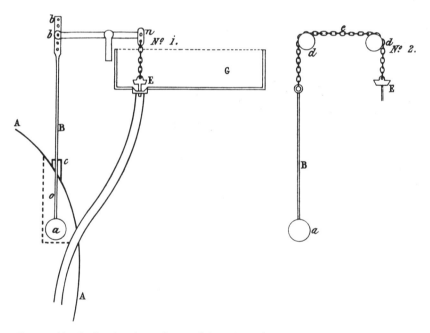

Figure 48 Boiler level regulator of S. T. Wood.

5. Final Acceptance as a Component of Steam Boilers

The fact that in 1784 the float regulator was protected for the second time in Sutton Thomas Wood's patent obviously does not establish Wood's priority; it does indicate, however, that the device was not yet commonly known. Boulton & Watt did not employ it, initially. James Watt's operating instructions for steam engines of 1779 show clearly that the boiler feed was regulated by hand. Its paragraph 64 ends with the characteristic sentence: "By a little attention, you will find the proper opening of the feeding cock for any rate of working."[160] A later set of instructions from approximately 1784 still refers to manual feed regulation: "The water in the boiler should be kept as nearly of the same height as possible; as carelessness in this point may cause the most sudden destruction of the boiler...."[161]

But soon afterward Boulton & Watt adopted the invention. An early witness to this is Georg von Reichenbach (1772-1826) who, in the service of his sovereign the Elector of the Palatinate, spent the summer of 1791 at the Soho works to study Watt's steam engine. A remarkably accurate freehand sketch of the engine that he secretly prepared does not show actually the float regulator, but his diary includes a very clear description of it:[162]

Nun ist auch ganz begreiflich, dass durch die Abrauchung sehr viel Wasser aus'm Dampfkössel verloren geht und dass das Wasser immer eine gewisse Höhe halten muss; folglich muss ein Mittel seyn, um dieses abrauchende Wasser zu ersetzen; weil es nun gefährlich ist, dass man durch Eröfnung eines Kranes mit der Hand das Wasser in dem Kössel regulieren muss, aus Ursache weil man nicht versichert ist, ob der Zulauf der Abnahme gleich ist, und am Ende der Kössel entweder ganz voll, oder ganz leer würde, wie es an denen alten Feuermaschienen öfters geschieht, so dachte Herr Watt darauf, das Wasser im Dampfkössel zu regulieren, ohne eine menschliche Hand daran zu bringen, welches das sicherste ist; die Einrichtung davon ist folgende: es schwimmen auf der Oberfläche des im Dampfkössel stehenden Wassers zwey wasserhaltende kupferne Kugeln, welche durch eine kleine Stange mit einander verbunden seyn, an welcher wieder ein kleines Stängelgen oben durch den Dampfkössel luftdicht geht; wenn nun das Wasser im Dampfkössel steigt, so werden die zwey Kugeln auch steigen, mit letzt benanntem Stangelgen, welches oben an einem sehr kleinen Hebelgen befestigt ist, an dessen entgegengesetzter Seite ein kleines über einem Rohr gehendes Fendil geschlossen würd, und den Eingang des Wassers verhindert, bey entgegengesetzter Richtung, würd es sich just umgekehrt verhalten, folglich behalt das Wasser immer eine Höhe.

Now it is quite understandable that by evaporation a great deal of water is lost from the boiler, and that the water must always be kept at a certain level; hence, there must be a means to replace this evaporated water. Now, since it is dangerous to regulate the water in the boiler by manually opening a valve, for the reason that one cannot be sure whether the inflow is equal to the consumption, so that in the end the boiler might either be quite full or quite empty, as often happens with the old "fire engines"; therefore, Mr. Watt thought out a way of regulating the water in the boiler without putting a human hand to it, which is the safest. The contrivance is the following: two watertight copper balls floating on the surface of the water in the boiler are connected by a small bar to which is attached another small bar which passes airtight through the boiler wall by means of packing. If now the water in the boiler rises, the two balls will also rise, together with the last-mentioned bar, which is connected on top with a very small lever; the opposite side of the lever will close a small valve in a pipe, preventing the entry of water; in the opposite direction, the action will be reversed; thus the water maintains one level at all times.

The float regulator by that time was part of the standard equipment of Watt's steam boilers. For Reichenbach, however, the device was evidently new. Dickinson and Jenkins have published three drawings from the Boulton and Watt Collection, showing float regulators dating from 1794 ("Tate's Boiler") and 1803 ("Coates' Engine" and "Radford Cotton Company"). They are similar to Wood's design, except that the 1803 devices are combined with steam-pressure regulators acting on flue dampers (Fig. 54).[163]

By the end of the 18th century the float regulator was universally accepted in English boiler construction. In the years following it was also introduced to the Continent, along with the steam engine itself.

VIII. Pressure Regulators

At about the same time, in the wake of the float regulators as it were, another class of regulating device appeared on steam boilers: the pressure regulators. They were actually not entirely new, for they had had an early predecessor in Denis Papin's safety valve.

1. The Safety Valve of Papin

Denis Papin (1647-1712), as a token of gratitude for his election into the Royal Society, published in 1681 the description of a pressure cooker in which the pressure was regulated by a weight-loaded valve (Fig. 49).[164] In 1707, Papin used the same device with a new purpose, namely as a safety valve on his high-pressure steam engine.[165] As early as 1717/18 the safety valve was a standard accessory of the steam engines built by Desaguliers.[166] Hereafter it appears on practically every steam engine, without special discussions in the texts: technology had accepted it as a conventional machine element.

Figure 50 shows the closed feedback loop. (The second, interior loop is not caused by any control action; rather it is thermodynamic in nature.) The actual steam pressure, converted by the area of the valve lid into an upward force, is compared with the desired pressure which is represented by the load on the valve. If the actual pressure is greater, the valve lid opens, the surplus steam will escape, and the pressure will fall.

The decision of whether this device deserves to be classified under *feedback control* depends on which third criterion one accepts. A *sensing element* can clearly be identified, but the functions of sensing, comparing, and control action are not physically separated: the valve plug fulfills all three.

The invention of the safety valve is unanimously ascribed to Papin; predecessors are not known. But it is difficult to prove that Papin had been aware of the closed control loop when he invented the device, as might be argued in the case of the thermostats of Drebbel and Becher.

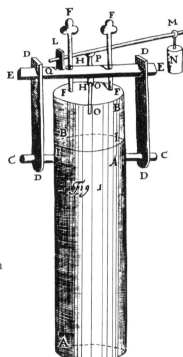

Figure 49 Papin's pressure cooker with pressure regulation.

2. Robert Delap

A more complicated pressure regulator is described in British patent No. 2302 of April 6, 1799, granted to Robert Delap of Banville, Ireland, on the subject of "Certain Oeconomical Boilers for sundry useful Purposes" (Fig. 51):[167]

... When these boilers are used to work steam engines..., the steam ought to issue at all times with equal force, for which purpose the regulator, Fig. 8, is added to the steam pipe. l, the beam; 2, a cylinder open at

84 PRESSURE REGULATORS

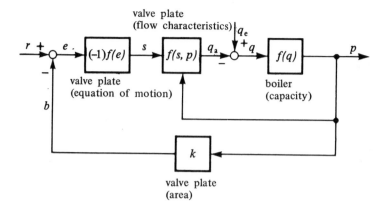

r   =   desired pressure (load on valve plate)
p   =   actual pressure
b   =   pressure force on valve plate
e   =   b − r

s   =   valve opening
q_a =   discharge of steam
q_e =   rate of evaporation
q   = $q_e - q_a$

Figure 50 Papin's pressure regulator, block diagram.

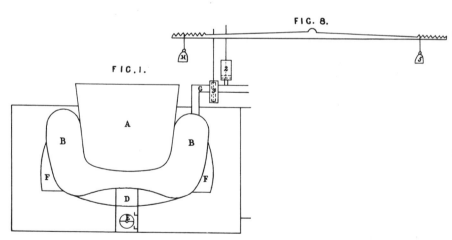

Figure 51 Pressure regulator of R. Delap.

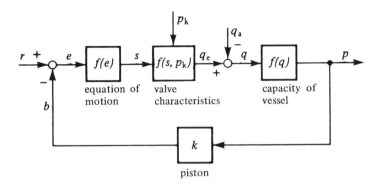

r = desired pressure: load on piston	s = valve opening
p = actual pressure	p_k = boiler pressure
b = pressure force on piston	q = flow rate of steam
e = $r - b$	

Figure 52 Pressure regulator of R. Delap, block diagram.

top, but close at bottom (except the small tube which connects it with the main steam pipe); in the cylinder is a piston well packed; 3, a sliding valve, or cock, etc.; 4, 5, two weights to be used as occasion requires. It operates thus: — When steam in equilibrium with the atmosphere is wanted, both weights 4 and 5 are taken off; the beam is then a true balance, having the piston valve, etc. attached to it. The engine is set to work by pushing down the valve or turning the cock, etc. which opens the steam passage; the piston is at the same time brought near to the bottom of the cylinder; they will remain in this state while the steam is in equilibrium with the atmosphere; but the instant it becomes stronger the piston is raised, and with it the valve or cock, etc., and thereby the steam passage is contracted, and thus the equilibrium will be restored, of course an equal motion in the machinery preserved. If stronger steam than the atmosphere is wanted, the weight 4 is put on, and the weight 5 off. If the reverse, the weight 5 is put on and the weight 4 off. The valve or cock, etc. must be placed between the boiler and this cylinder.

The mechanism regulates the pressure of the steam entering the engine. The actual pressure is sensed by the piston *2* which in turn initiates corrective action by adjusting the throttling valve *3* which is located upstream. The desired pressure is determined by sliding the weights *4* (misprinted as *H*) and *5*. The block diagram (Fig. 52) shows the closed feedback loop.

The invention suffers from a certain shortcoming which would have to be removed before it could function adequately. If after a period of

86   PRESSURE REGULATORS

equilibrium the steam pressure has risen higher than desired, the resulting force acting upon the piston 2 will be upward. Since this force is unresisted, except for friction, the piston will be driven upward until it meets some resistance, namely until the valve 3 is closed. Now the pressure will promptly drop, causing the piston to fall down until the valve is wide open. The pressure will rise again, and the whole cycle will repeat itself. The valve will always be either fully open or closed: the system thus provides only on-off control. The small time constants of pressure control systems make it impossible for even an approximation of equilibrium to establish itself; the system therefore fails in its purpose that steam should "issue at all times with equal force."

The difficulty could easily be removed by converting the on-off regulator so as to provide proportional control: by loading the piston with a helical spring, the position of the piston can be made a measure of steam pressure.

3. Matthew Murray

Three months after Delap (July 16, 1799) Matthew Murray of Leeds received a patent on a pressure regulator that avoids this pitfall (Fig. 53). The device is specified as follows:[168]

> First, in respect to principles, I cause the steam contained in the boiler ... to act on the intensity of the fire in such a manner, that when the steam is increased in the boiler beyond its proper density, the fire will in proportion decrease in its intensity or heat, so as to keep a proportion between the density of the steam and the draft or consumption of fuel.
> Fig. 1 and 2, plan and section of a boiler. A, a small cylinder upon the boiler, in which is fitted the piston and rack B,B, which are made to move freely up and down. C is a small wheel upon the shaft or spindle D, and works with the rack B. E is a damper fixed to the end of the spindle D in the chimney F, where it has free liberty to turn round. G is a circular cone fixed upon the spindle D, from which the weight H is suspended by a small chain. I is a pointer upon the spindle D, and is made to go round the dividing scale K. Now, as the steam increases in the boiler beyond what is necessary, it will press up the piston and rack B, which will turn the wheel C and shut the damper E, at the same time it will wind up the weight H, by which means the draught of the chimney is suspended, and the further consumption of the burning coal is stopped till the superfluous steam is wrought out of the boiler, while the divided scale shews the density of the steam and regulates the attention of the fireman.

A crucial feature is the spiral-shaped cam G fastened to the axle D. Rolled up over the circumference of the cam and suspended from it is a chain carrying a weight H. The cam G, the chain, and the weight H produce

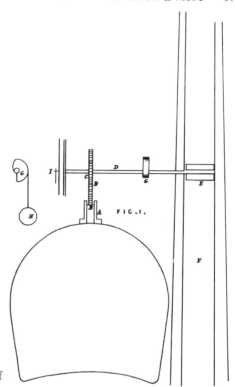

Figure 53 Pressure regulator of M. Murray.

upon the axle D a torque which increases with rising pressure, counteracting the torque produced by the steam pressure through the rack and pinion. The damper E thus rotates in proportion to the boiler pressure, giving the system the characteristics of proportional control. The effectiveness of the regulator is impaired by a source of error in the method of pressure measurement. The upward force on the piston, taken as a measure of steam pressure, is resisted by an unpredictable frictional component which is not negligible since the piston must be well sealed within the cylinder A.

4. Boulton & Watt

A third pressure regulator appears first in 1803 on two drawings made by the Boulton & Watt firm for engines ordered by Coates & Co. and by the Radford Cotton Co. (Fig. 54a and b).[169] Its arrangement clearly shows its derivation from the float regulator, with which, incidentally, it is also

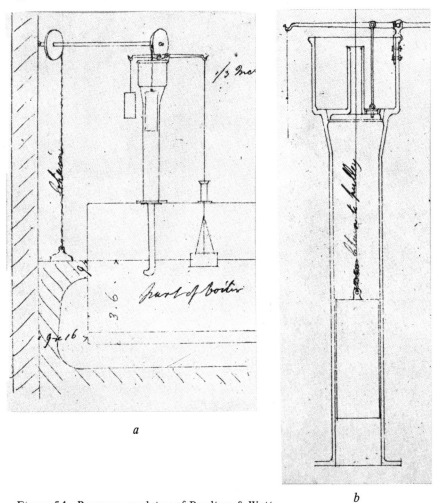

Figure 54 Pressure regulator of Boulton & Watt.

directly combined (Fig. 54a: the float of the level regulator is on the right). The steam pressure in the boiler is measured by means of the water column in a vertical pipe passing through the boiler wall; the bottom of the pipe is immersed in the water of the boiler, while its top is open to the atmosphere. The difference between the water level in the pipe and that of the boiler in general is the measure of the boiler pressure (above atmospheric); Watt's boilers operated at only very slight overpressures. The height of the water column is sensed by a cylindrical float (Fig. 54b) which is connected, through pulleys and chains, to a damper in the flue which closes with increasing pressure. Thus the control loop is closed. The desired pressure is determined by the length of the chain. This regulator seems to have proven practical. It was employed in 1820 by William Brunton, for example, in a complicated system of furnace regulation which worked quite successfully, as he testified before a committee of the House of Commons in 1822.[170] A pressure regulator of this type is also employed on a steam engine that was operated in Paris about 1830.[171]

IX. Feedback Control on Mills

In addition to the float regulators and the thermostats, a diverse group of regulators devised by the millwrights forms a third family in the evolution of feedback mechanisms; these have a special historic significance, since they are the direct antecedents of James Watt's famous centrifugal governor.

1. The Mill-Hopper

First to be discussed is the mill-hopper, a mechanism that has been regarded by some authors as the oldest feedback device;[172] it will be seen that the term *feedback device* does not apply to it any more than the attribute *oldest*. The mill-hopper (also known as *shoe-and-clapper grain feeder, shock distributor,* or in French, *baille-blé*) supplies the millstones with grain. In details of design it may differ widely, but in principle it is always a loosely suspended hopper (Fig. 55) from which the grain is poured into the opening in the center of the upper, rotating millstone. Attached to this rotating millstone is some eccentric projection, which periodically during each revolution knocks against the hopper in order to keep the grain flowing. The rate of flow of the grain can be adjusted by changing the inclination of the hopper.

A crude representation of this idea is already shown in the *Hortus deliciarum* (ca. 1200) by Herrad von Landsperg;[173] Ramelli (1588) has several more refined versions of the device.[174] By the beginning of the 18th century the final form,[175] which was retained unchanged into the

6 - 7   hopper, suspended by ropes
k   rotating millstone
i   fixed millstone
a   rod attached to hopper,
        struck periodically by
        eccentric projections on
        the rotating millstone.

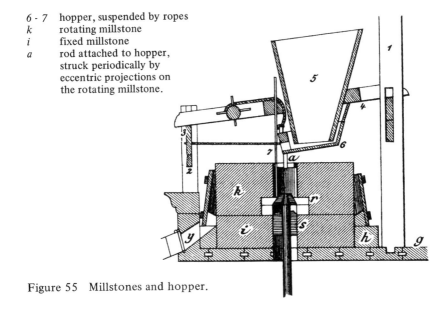

Figure 55   Millstones and hopper.

20th century had established itself.[176] The mill-hopper is not documented earlier than Herrad von Landsperg; with regard to age it is therefore no rival for the Hellenistic float regulator.

The claim that the mill-hopper is a speed regulator is based on the following argument:[172] When the speed of the mill increases, for example due to rising wind, the hopper receives more knocks, and more grain is poured upon the millstones. The resulting increase in load resistance, it is argued, will slow down the rotation of the mill, thereby completing a closed feedback loop. That it would not be a very effective regulator is readily conceded, but such an objection would be irrelevant. The question whether this should be classified as feedback control must be decided by qualitative criteria, whereas effectiveness is a matter of degree.

Although at first this argument sounds plausible, the following objections come to mind:

(A) It was not the purpose of the mill-hopper to regulate the mill speed. According to a representative 19th-century textbook,[177] the mill-hopper had to satisfy only the following requirements: The flow of grain had to be manually adjustable; the flow of grain had to increase with increasing speed, and vice versa; and the flow of grain had to be continuous. There is no allusion to speed regulation as such. Indeed, no other sources are known assigning this function to the mill-hopper. If the device neither did

92 FEEDBACK CONTROL ON MILLS

effectively regulate the speed, nor was intended to, then it seems unjustified to classify it as a regulator.

(B) One must distinguish between the action of the hopper and that of the millstones. The millstones' capacity of milling grain was directly related to the mill's speed. At each speed there was a definite flow rate of grain that the millstones could process. This quantity increased with speed. The hopper had only the purpose of supplying the correct rate of grain at every speed, but it was not capable of affecting the speed of the mill by changing the rate of grain to be processed by the stones. In other words, the hopper was not able to influence the stones; instead, it received its orders from them. If the hopper supplied more grain than required, the grain would pile up; if it supplied too little, the stones would run dry and catch fire. If rising speed is accompanied by an increase of milling rate and hence an increase of load torque which will resist further speed increases, then this effect is the result of the various kinds of friction, especially that between the millstones, and not of the action of the grain-hopper.

(C) It is not possible to identify in this system any particular element with the specific function of sensing the alleged controlled variable, speed. The signal that closes the causal loop, the resisting frictional torque, cannot in all reason be interpreted as a deliberate measure of mill speed.

If not a feedback control system, what then is this system? The block diagram (Fig. 56) shows the closed loop. The driving torque, supplied by the wind, sets the mill into motion after overcoming the resistance of inertia. This leads to an increase in friction (millstones, transmissions, bearings). As soon as the resultant braking torque has become equal to the driving torque, the speed of the mill will remain constant. If afterwards the wind should increase further, speed as well as friction would rise again, until another, greater, equilibrium speed has established itself. The velocity of the mill then is by no means constant, although it will respond to disturbances with changes of only limited magnitude.

m_a = driving torque (due to wind)
m_b = braking torque
m_r = $m_a - m_b$
n = speed

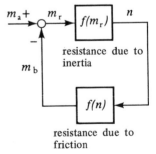

Figure 56 Milling process, block diagram.

The mill-hopper thus belongs to the class of systems with self-regulation which due to their particular dynamic characteristics can be formally considered closed loops,[10,11] even though conceptionally they have nothing to do with feedback control (de Latil[172] has described numerous such systems). The property of self-regulation is inherent to them, it is not the result of deliberate design, as would be the case if the comparator, the feedback path, or the sensing device could be identified as physically distinct elements. All this makes it clear that the mill-hopper has no significant place in the history of feedback control.

2. The Fan-Tail

A number of ingenious feedback mechanisms were invented by English and Scottish millwrights of the 18th century. The study of these inventions is subject to a peculiar limitation. In practicing his profession, a millwright had little incentive to leave written records of his work. The only occasion to put down an idea in writing was the patent application. The main source for the following chapter then are the specifications of old British patents. Biographical details about individual inventors and sometimes also the fate of the inventions remain unknown.

The oldest of these devices, the *fan-tail,* was destined for an unusually successful career, especially in Britain, as an accessory to windmills. Consisting of a small wind wheel mounted at right angles to the main wheel, it was designed for the purpose to point the mill continuously and automatically into the wind. Figure 57a shows a typical design from the 19th century. The invention of the fan-tail is often falsely credited to Andrew Meikle, 1750.[178]

It had actually been patented as early as in 1745 (British Patent No. 615) by the blacksmith Edmund Lee of Brock Mill, Wigan, Lancaster under the title "Self-regulating Wind Machine."[179] In 1747 the same Lee also received a patent (No. 242) of the States of Holland on "the invention of a wind machine which—by its regular working—will excell all wind-, flour-, and other mills."[180] The Dutch patent reveals no further details, but it probably covered the same invention. Figure 58 shows the original drawing from Lee's patent specification, which was accompanied by the following explanation:

A, the case of the Machine, *B*, the sails, *C*, the Regulating Barr passing thro' the center of the originall axes, *D*, the Chains from the Barr to the Sails, *E*, the Back Sails which keep the machine Constantly in the wind, *F*, the weight which regulates the Sails according to the winds force, *G*, the Travelling wheel which moves on planks round the machine, *H*, the Regulator to which the weight is Fixed.

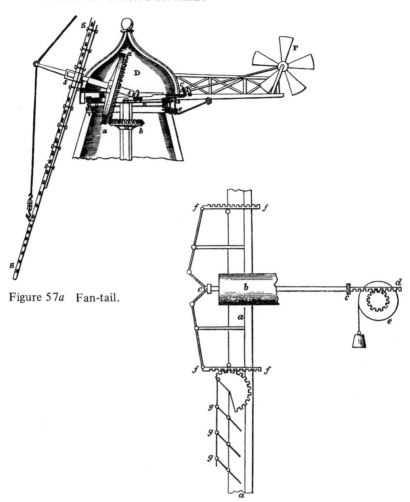

Figure 57a Fan-tail.

Figure 57b W. Cubitt's *patent sails*.

Lee describes here two different inventions at once:
(a) the fan-tail, termed by him *backsails*, whose function it is to keep the mill continuously directed into the wind;
(b) a mechanism for automatically varying the angle of attack of the windmill sails. The sails can be tilted about their lengthwise axis and are held in a rest position (with a maximum angle of attack) for low wind speeds by a counterpoise (lower right). With increasing wind, when the force upon the sails exceeds that of the counterpoise, the

sails bend backward, with the result that excessive speeds are avoided. This mechanism will be treated in greater detail below.

Lee's drawing, in spite of some errors in perspective, shows clearly the functioning of the fan-tail. Only the cap of the tower mill shown can rotate. The fan-tail mechanism is rigidly attached to this cap. The wind wheel E, through the gear train of a transmission, drives the gear G which engages a circular rack anchored on the ground. If the mill is not facing the wind, the wind will strike the tail wheel, causing it to rotate, slowly turning the cap of the mill into the wind. The motion will stop when the tail wheel stands parallel to the wind. The result of the tail wheel's rotation therefore is to eliminate its own cause, that of not being parallel to the wind; it acts in the pattern of a closed loop (Fig. 59). In contrast to all the regulators discussed so far, where the command signal was constant, the controlled variable on the fan-tail (the direction of the mill) must be kept equal to a continuously changing command signal (the direction of the wind); it is therefore a *follow-up,* or *servo* system.

The history of the fan-tail's introduction into practice is not known in detail. Both inventions of Edmund Lee reappear in 1773 in an English patent (No. 1041 "Smelting Furnace, etc.") of John Barber.[181] The heart of this patent is a blast furnace with a wind-powered blower. Presumably Barber did not intend to have the idea of the fan-tail protected as being his own invention, but was using it simply as something generally known. The fan-tail shown by Barber is of a more advanced design than that of Lee: it is directly attached to the mill-cap in a manner that became customary in the 19th century. Rex Wailes has expressed the opinion that the fan-tail could have come into general use only after the technology of the iron foundry had progressed far enough so that the various gears required in the drive train could be cast in iron.[182] Fairbairn wrote in 1861 that "modern" windmills were now equipped with the fan-tail.[183] The great many 19th century windmills shown in Rex Wailes' book *The English Windmill* testify to the general use of the fan-tail in England.[184]

On the European continent the fan-tail was by no means universally accepted. It is accurately described in a German manual on windmills of 1850,[185] and indeed it was successfully introduced to northern and western Germany as well as to Denmark;[186] in France, however, and in the rest of Europe reportedly it was never accepted.[187]

3. Self-regulating Windmill Sails

Among the mechanisms designed to regulate the speed of windmills, the oldest and simplest belong to the class of self-regulating windmill sails.

96   FEEDBACK CONTROL ON MILLS

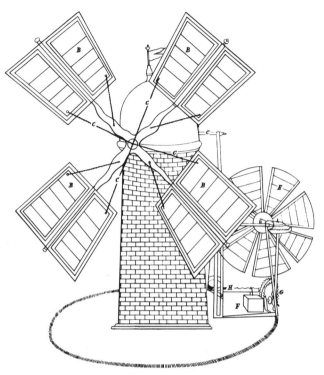

Figure 58 E. Lee's patent drawing of a windmill with fan-tail and self-regulating sails.

This invention was proposed in several forms in the 18th century, and achieved a certain amount of popularity in the 19th century.[188] From the standpoint of feedback control it is a borderline case.

To study its function, we return to Lee's drawing (Fig. 58): each of the four windmill sails is pivoted so that it can be tilted around its lengthwise axis in such a manner that the center of pressure of the impinging wind will lie behind the axis. A set of chains C running through the hollow main shaft of the mill, connect the trailing edges of the sails with a counterpoise F in the rear. The weight draws the trailing edges of the sails forward toward the wind, although in a flexible manner. With increasing wind the sails will be pitched farther and farther backward, projecting a smaller area into the wind, so that the force driving the mill will hardly increase in spite of the growing wind; the result resembles speed regulation.

This invention first appeared, as far as known, in Edmund Lee's patent of 1745. Modifying this idea, Andrew Meikle, the famous Scottish millwright, in 1772 proposed to substitute springs (the "spring sail") for the

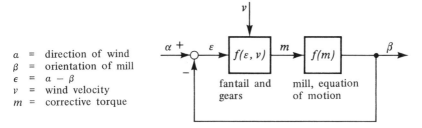

a = direction of wind
β = orientation of mill
ε = α − β
ν = wind velocity
m = corrective torque

Figure 59 Fan-tail, block diagram.

counterpoise. He did not apply for a patent; his design has only survived in a hand sketch, now in the library of the Royal Society, which he had sent to John Smeaton.[189]

Lee's arrangement found two further imitators: John Barber in his above-mentioned patent (No. 1041) of 1773 modified the shape of the wind wheel.[181] He replaced the four large sails with a large number of small disks, arranged on a circular frame the size of an ordinary wind wheel. Each of these disks was pivoted and held into the wind by a counterpoise, similar to Lee's arrangement. An even closer imitation is the patent of Benjamin Heame (No. 1588) of 1787.[190] It is distinguished from Lee's model only by greater thoroughness of execution: while Lee's sketch barely indicates the principle, Heame furnishes an exact drawing.

This development is concluded in 1807 with a patent (No. 3041) of William Cubitt.[191] Here the wind sails are divided up as movable louvers which are, as before, connected to a common counterpoise by means of linkages. In the 19th century Cubitt's *patent-sails* were considered standard equipment of up-to-date windmills (Fig. 57b).[192]

It remains to be discussed whether the self-regulating windmill sails indeed represent feedback control. Their purpose is to maintain a constant speed of the mill in spite of any disturbance caused by changing winds. They apparently fulfilled this duty, perhaps not with the precision of modern control equipment, but sufficiently so for several generations of millers. The flexible windmill sail, adopting a new equilibrium position for each wind velocity, functions, as can be shown, in a closed loop (Fig. 60). Two of our criteria for feedback control are thus fulfilled. But in addition we require of any genuine speed regulation that somehow or other it will actually measure the controlled variable, speed. Such a sensing element cannot be identified here. The speed of the mill does not in any way react back upon the control system. The corrective action is causally independent of the controlled variable. The system hence does not represent true feedback control.

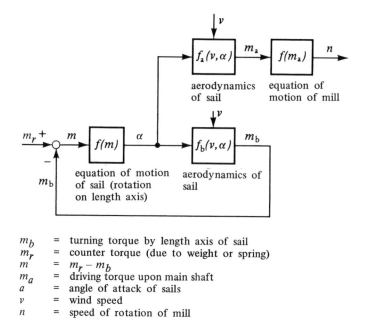

m_b = turning torque by length axis of sail
m_r = counter torque (due to weight or spring)
m = $m_r - m_b$
m_a = driving torque upon main shaft
a = angle of attack of sails
v = wind speed
n = speed of rotation of mill

Figure 60 Self-regulating windmill sails, block diagram.

How then is the effect of speed regulation produced, if it is not due to feedback control? The counterpoises (or the force of the springs) exert a torque upon the windmill sails which tends to turn them forward, increasing their angle of attack (the angle between sail profile and direction of wind). This torque, by design, is virtually independent of the angle of attack. On the other hand, the wind generates a torque in the opposite direction, twisting the sails backward into positions of decreased angle of attack. This torque increases with growing wind speed but decreases with diminishing angles of attack (it becomes zero when the sail is parallel to the wind).

In a sudden gust of wind, for example, the increased torque will tilt the sails backward, but with the decreasing angle of attack this torque itself will decrease as well until it is balanced by the constant countertorque. At this particular angle of attack an equilibrium will establish itself. The action can be represented as a closed loop (Fig. 60): the signal that corresponds to the controlled variable is the torque due to the wind turning the sails around their length axis; the system is an example of self-regulation.

The speed of the mill is not part of the closed loop. It lies outside of the loop, being a dependent variable of the angle of attack. The propulsive force upon the sail, perpendicular to the direction of the wind, decreases with the angle of attack but increases with increasing wind velocity. Owing to the characteristic property of our system that the angle of attack decreases with rising wind, the force driving the mill will remain fairly constant over a certain range of wind speeds, and thus the system provides a semblance of speed control.

4. Sensing the Speed of Rotation

A prerequisite for designing a speed regulator is to have a sensing device that will express speed in terms of some physical quantity which in turn is capable to initiate corrective action upon the system. The following three patents, dating from 1785 to 1789, describe devices for controlling the gap between millstones (commonly known as *lift-tenters*). These do not actually employ feedback, even though they have been referred to as *governors*. Their significance in this context is only that they represent the first practical arrangements for sensing the speed of rotation. All three cases address themselves to the same basic problem: whenever, due to rising speed, the flow rate of grain fed into the milling process is increased, the stones have a tendency to separate from each other. Since the widening gap leads to a lower quality of flour, this tendency must be opposed. Some speed-sensing device is needed that will generate a force, proportional to the speed, that will press the millstones together.[193]

(a) Centrifugal Fan with Baffle

A solution to this problem was first patented in 1785 by Robert Hilton (British Patent No. 1484).[194] It covers "certain Improvements in Windmills, consisting of a Method ... of Raising and Falling the Acting Mill Stones in the Mill, so as to cause them to Grind to much greater Advantage by the Motion of the Mill according to the different Force and Pressure of the Wind upon the Sails thereof..."; in addition, it describes a freewheeling mechanism for the mill action and an arrangement for the mechanical unfurling of windmill sails, neither of which is relevant here. Figure 61 shows Hilton's drawing of his speed-sensing device. The rotating stone P (the fixed stone located below it is not shown) can be lifted or lowered by an arrangement of levers. The mill speed is sensed by a centrifugal fan AB driven from the mill shaft. The flow of air discharging from the fan is blocked by a movable baffle in the outlet shroud. This baffle is connected, through the rod F and the shaft D, to a pulley E driving a rope

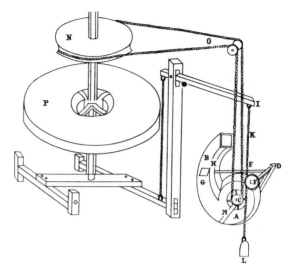

Figure 61 Hilton's speed sensing device.

K which is attached to the lever I that lifts or lowers the millstone. With rising mill speed the discharge of the fan, and hence the air pressure in front of the baffle, will increase. The baffle is forced backward and, through the action of rope, pulley, and levers, causes the millstone to be lowered. Similar to the mill-hopper, this device has the effect of self-regulation upon the speed of the mill, although only to a small extent. As the stones are pressed together more strongly, the friction increases and the motion is braked. But this was not the intention of the design. By all indications, the purpose of the mechanism was solely to maintain a constant quality of flour by controlling the gap between the stones.

This method of sensing speed might have proven quite adequate for speed regulation; it might have found some practical acceptance, had it not met a far superior rival in the centrifugal pendulum.

(b) Centrifugal Pendulums

In 1787 Thomas Mead took out a patent (British Patent No. 1628) on a new solution of the same problem to which Hilton had addressed himself.[195] His basic conception of the lift-tenter was the same, but in addition to a few refinements in design he described an important innovation: the speed of rotation was to be measured in terms of the centrifugal motion of a revolving pendulum (Fig. 62). By a simple arrangement of sleeves and levers Mead linearized this motion in order to employ it for

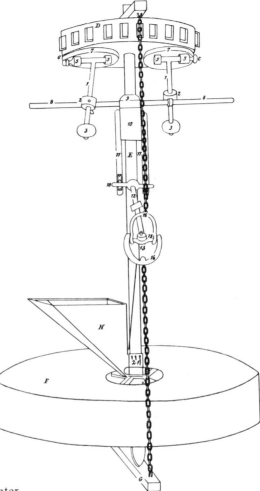

Figure 62 Mead's lift-tenter.

pressing the millstones together. But the patent also included an invention which went a step further: the displacement of the pendulum was so arranged as to react back on the system, thereby closing the loop to form a genuine speed regulator.

Mead described his invention as "a regulator on a new principle," but it is not clear whether it was actually he who was the inventor of speed regulation by means of centrifugal pendulums. In 1825, Dr. Alderson, president of the Hull Mechanics' Institute, stated that his "late friend

Mead, who, long before Mr. Watt had adopted the plan [i.e., the centrifugal regulator] to the steam engine, had regulated the mill-sails in this neighbourhood upon that precise principle, and which continued to be so regulated to this day."[196] James Watt himself later carried on some correspondence in order to establish the facts of this matter; his informants thought the device had been in use prior to the date of Mead's patent, but they could not establish verifiable evidence.[197]

In another, much simpler, application we encounter the centrifugal pendulum as early as the 15th century: Weights suspended from chains serve in various crank drives in the role of a flywheel. Arrangements of this type are found in an anonymous manuscript *"From the Time of the Hussite Wars"* ca. 1430, in the *Mittelalterliches Hausbuch* of ca. 1480, in the writings of Francesco de Giorgio Martini, ca. 1482-1501, and in those of other Renaissance engineers.[198] The centrifugal pendulum of Christiaan Huygens also belongs into this class.[199] Although these devices have occasionally been referred to as regulators or governors, they are not true anticipations of James Watt's regulator, but merely special cases of flywheels.

In 1789 the centrifugal pendulum reappeared in a patent (British Patent No. 1706) of Stephen Hooper, who applied it to the same task as Mead.[200] His arrangement of a lift-tenter is so closely modeled after that of Mead, that it alone would hardly have been worthy of a patent (compare Figs. 63 and 62). The main part of that patent, however, concerned an invention of somewhat greater originality, a speed regulator for mills, which will be discussed further below.

The lift-tenter achieved a certain popularity among British millwrights, as is demonstrated by numerous 19th-century mills in Rex Wailes' *The British Windmill* which are equipped with this device.

But the true destiny of the centrifugal pendulum lay not in the windmill but in the classical governor of the steam engine.

5. Speed Regulation

(a) Mead's Regulator

Mead's patent of 1787 describes three applications of the centrifugal pendulum.[195] Apart from the lift-tenter, these are an automatic transmission and a speed regulator for windmills in which the centrifugal pendulum determines the sail area of the windmill.

In the automatic transmission the gear ratio is continuously adjustable; it is controlled from the input side so as to be always inversely proportional to the input speed. As a result, the output speed should remain constant in

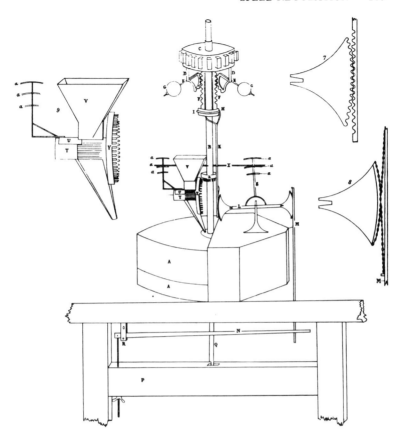

Figure 63 Hooper's lift-tenter.

spite of fluctuations in the speed of the drive. Since the controlled variable, the output speed, cannot react back upon the system, there is no closed loop, and hence no feedback. The system is an example of feed-forward, i.e., a special case of open-loop control.

The other system, the speed regulator, is referred to in the patent specification as "a Regulator on a new Principle for Wind and other Mills, for the better and more regular Furling and Unfurling the Sails on Windmills without the constant Attendance of a man." Figure 64 shows this invention in Mead's patent drawing, in which allowances must be made for some minor imperfections.

The centrifugal motion of the balls *30* is transmitted, through the ropes *46* and *44* and pullies *32* and *45*, to the double sleeve *33* and *34*, which

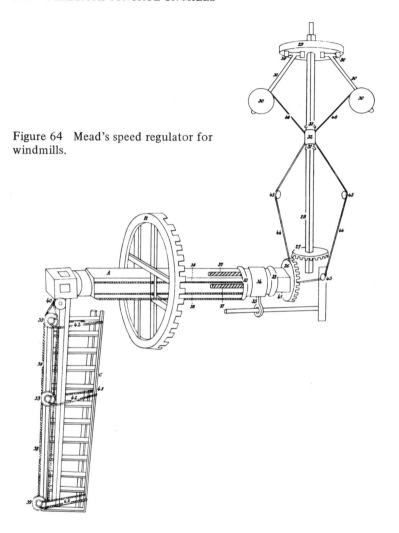

Figure 64 Mead's speed regulator for windmills.

proportionately slides forth and back on the main mill shaft *A*. The ropes *44* are attached (not shown) to the outer sleeve *34* which is prevented from rotating by the guide *35*, while the inner sleeve *33* rotates with the main shaft *A*. Helical springs *37* pull the sleeves forward toward the mill wheel. The axial motion of the sleeves *33* and *34* determines the sail area by means of two ropes (the rope *38* runs between the pulleys *39* and a pulley near *41*, not shown); the rope *42* runs between the pulley *39* and the beam *43* on which the canvas of the windmill sails is coiled up.

The device works as follows: with rising speed the centrifugal weights swing outward and, through the pull of the rope, force the sleeve to slide to the right, i.e., toward the rear of the mill. As a result of this the canvas is rolled up, i.e., the sail area is reduced. Figure 65 shows the block diagram with the feedback loop.

The comparison between the actual speed (controlled variable) and desired speed (command) is carried out as a comparison of forces upon the sleeve 34, which is acted upon by the opposing forces of the springs and the centrifugal force. The outcome of the comparison (i.e. the deviation from the desired value) is expressed in the axial position of the sleeve.

The system seems to be designed correctly enough, although we miss some of the accessories required in practical operation. The desired speed (i.e. the precompression of the helical springs) can be changed only when the mill is stopped. Also, no provisions are made to neutralize the regulator in order to enable the operator to adjust the sail area by hand. Finally, it seems quite difficult to change the sensitivity of the system (the ratio between a given speed deviation and the resulting change in sail area); this would require changing either the spring constants of the springs or the transmission ratios of the pulley drives. Regarding the practical fate of the invention we have only the testimony of Dr. Alderson (p. 101), who said that it was used to regulate windmill sails in the neighborhood of Hull as late as the 1820s.

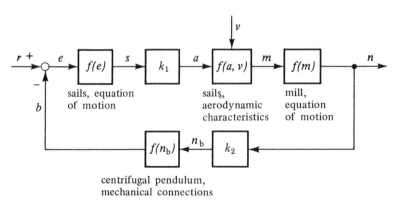

r = desired speed (spring force)
n = speed of mill
n_b = speed of centrif. pendulum
b = centrifugal force (transmitted)
s = axial position of sleeve
a = sail area
v = wind speed

Figure 65 Mead's speed regulator, block diagram.

(b) Hooper's Regulator

Hooper's patent of 1789 contains, beside a not very original lift-tenter, an invention of a "Certain new-constructed Machinery for regulating the Power and Motion of Wind and other Mills...",[200] which is commonly referred to as *roller reefing sails*.[201]

Basically, the invention deals with a method for manually varying the sail area during the operation of the mill (Fig. 66). Each sail consists of a rigid frame, composed of the main beam *B*, the crossbars *E*, and the necessary longitudinal bars (unmarked), and a second loose frame, made up of the longitudinal bars *G* and the rollers *F*, the latter free to slide lengthwise relative to the rigid frame. The individual sections of canvas *D* are attached on one side to the fixed rods *E* and on the other to rollers *F*, which slide with the detached frame. The loose frame can be moved radially on the windwheel from the inside of the mill by means of the handwheel *Q* through the rack-and-pinion drive *HIKLMOP* and the clutch *MN*. When it slides outward, toward the tips of the sails, the rollers *F* are rotated by some appropriate device (Fig. 66, top, shows several possibilities: rack and pinion, rope and pulley, and so on) so as to roll up canvas, thus decreasing the sail area. Conversely, when the loose frame slides toward the hub, the sail area is correspondingly increased.

How this arrangement acquires the character of feedback control is described by Hooper as follows:[200]

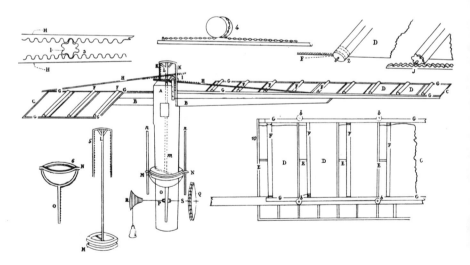

Figure 66 Hooper's speed regulator.

These sails will attend themselves when the Mill is at work by a weight connected with the extreme ends of the Rods *G.G.G.G.* so that when the weight by the velocity of the vanes or sweeps is forced out so as to overpower a counterweight that is fastened to a rope which works round a spiral wheel *R* fixed on the cross shaft *S* the sails roll up in proportion to such velocity and when the velocity abates the counterweight overpowers the weight on the vanes or sweeps and spreads the Sails.

Attached to the outer end of the loose frame are fly weights (not shown) which are balanced by a counterpoise inside the mill (below *R*) in such manner that they will take a particular radial position along the sail with each speed, approaching the tip with increasing speed. Thus, with rising speed the sail area is reduced; the cause of the speed increase is removed, and the feedback loop closed.

Conceptionally, then, this is a speed regulator. Here again the similarity between Hooper's and Mead's patents is obvious. Apart from differences in structural detail only the fly weights are moved outward from the interior of the mill to the tips of the sails. This not only creates additional problems for the strength of the structure; it also makes important parts of the control system so inaccessible that it would be difficult indeed to adjust this regulator for smooth operation. If Mead's superior system had no notable success, it seems plausible that Hooper's *roller reefing sails* were also not used for the purpose of automatic speed regulation.

The problem of speed regulation continued to occupy British millwrights: a British patent (No. 2782) of 1804 by John Bywater included a speed-control system equivalent to Mead's (for details, see p. 103). But little evidence is available to show that such devices were used in practice. Since 18th-century millwrighting is otherwise well documented, this seems to indicate that regulators of this type did not succeed. Indeed they held little promise: if the wind was too strong, the simpler and cheaper patent sails or spring sails were more reliable; if the wind was too weak, none of these systems could offer any help.

6. The Millwrights: Summary

This chapter, which is based primarily on the patent literature, demonstrates the abundance and diversity of inventions involving the concept of feedback that must be credited to British millwrights of the 18th century. Considering that in a purely practical profession such as millwrighting a certain proportion of all inventions was doubtless never patented or published in some other form, we may assume that feedback devices may actually have been even more widely used on mills. It is not surprising, however, that it was the millwrights who first learned to think in terms of this concept, for this professional group was distinguished by characteristics by which it exerted great influence on the subsequent evolution of mechanical engineering.[202]

The millwright was in a unique position: as a craftsman he had to combine the skills of the blacksmith, the mechanic, and the carpenter; but at the same time he was continuously challenged by the theoretical problems connected with wind and water wheels. Since his profession forced him to travel, news about the latest developments in his art would reach him quickly, and he was exposed to the intellectual currents of the time in the broader sense as well. He thus represents the link between the traditional craftsman and the scientifically trained engineer.

If, however, the millwrights as a group were unconventional, enterprising, and willing to experiment, they may occasionally have lacked earnestness and circumspection. James Watt, in an outburst of impatience, characterized them thus: "There is no end of millwrights once you give them leave to set about what they call machinery:—they have multiplied wheels upon wheels until it has now almost as many as an orrery."[203]

Both the virtues and the shortcomings of the millwrights are reflected in their regulating devices. A new idea was grasped with enthusiasm and imagination, but it was not always cultivated to the stage of maturity. It was only in another field, the steam engine, that the idea of feedback control became historically effective.

X. The Speed Regulation of the Steam Engine

1. James Watt's Centrifugal Governor

The feedback devices described so far either had remained obscure in unread books and in the shops of solitary inventors, or, when serving on actual machines, played an inconspicuous role. It was only by a detour through the field of the steam engine that by the turn of the 19th century the concept reached the consciousness of the engineering world. The Boulton-Watt engine, admired as a sensation, quickly spread over Europe. On it the attention was focused upon the centrifugal governor with its dramatically revolving flyweights (Fig. 1), impressively demonstrating the action of feedback. It is still widely believed that the steam-engine governor is the oldest feedback device, and that James Watt had not only invented but also patented it. While both errors are easily refuted, we are still not able to reconstruct the history of this invention in all desired completeness.

The centrifugal governor (Fig. 2) is based upon the rotating pendulum adopted from the millwrights, where the flyweights perform a centrifugal motion depending on the speed of revolution. By appropriate mechanical elements this motion is transmitted upon the inlet valve of the steam engine so that by throttling the steam flow the speed is reduced. A more extensive discussion of this action has already been given in the Introduction (p. 2).

A prerequisite for the invention of the centrifugal governor was of course a steam engine with a rotary output motion; this reached industrial maturity only in 1783-84. The older reciprocating engine, used to drive water pumps, had been regulated by the *cataract,* a device of archaic appearance that deserves to be mentioned briefly. It consists (Fig. 67) of the pivoted lever U carrying the vessel Y which receives a slow but steady flow of water from the faucet Z. When the vessel is empty, the left arm of the lever U points downward; when it is filled, the center of gravity of the arrangement is shifted until the balance of the lever is upset: the vessel drops over to the right, forcing the left arm of the lever upward. By this the vessel is emptied, returning the whole mechanism to its starting position. The lever is linked with the valve gear of the engine (*18-22*): each deflection of the lever triggers a work stroke of the engine. The engine speed is therefore determined by the flow rate of water through the faucet.[204] The cataract originated at the pumping engines of Cornwall coal mines; it was adopted in 1767 by John Smeaton[205] and in 1777 by Boulton & Watt.[206] In its similarity with ancient water clocks it forms a strange contrast with the centrifugal governor from which *cybernetics* borrowed its name.

Watt's first rotating steam engine was sold in 1783, others followed in 1784.[207] The main incentive for its development must have been the hope to introduce steam power into milling. Even before this engine was fully matured, Boulton & Watt became involved in the construction of a large and progressive steam mill in London. As supervisor of its construction and operation they hired in 1884 the 23-year-old millwright John Rennie (1761-1821) who had just completed his apprenticeship under Andrew Meikle and who was to become famous as a builder of bridges.[208] Early in 1786 the "Albion Mill" began its operation with one steam engine.[209]

The earliest document related to the introduction of the centrifugal governor is a letter dated May 28, 1788, from Matthew Boulton in London, who in this period spent a good deal of time at the Albion Mill, addressed to James Watt at the factory in Birmingham.[210] In this letter Boulton reports on various novelties that he had seen here, including a mechanism "for regulating the pressure or distance of the top mill stone from the Bed stone in such a manner that the faster the engine goes the lower or closer it grinds and when the engine stops the top stone rises up and I think the principal advantage of this invention is in making it easy to set the engine to work because the top stone cannot press upon the lower stone untill the mill is in full motion; this is produced by the centrifugal force of 2 lead weights which rise up horizontal when in motion and fall down when ye motion is decreased, by which means they act on a lever that is divided as 30 to 1, but to explain it requires a drawing."[211]

Boulton is referring to nothing else than the lift-tenter, patented by Mead in the year before and now installed, presumably by Rennie, in the

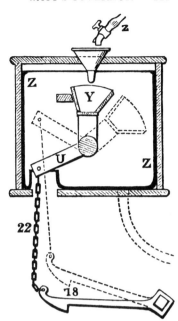

Figure 67   Cataract.

Albion Mill. Boulton's formulations make it clear that he was seeing the device for the first time. Half a year later, on November 8, 1788, an engineering drawing entitled "Centrifugal Speed Regulator" was finished in Watt's drafting room and taken to the shop by John Southern, Watt's assistant. In the following weeks John Southern was occupied several more times with the governor. On December 13 he prepared a new drawing from which the regulator on the famous "Lap" engine, now in the Science Museum, South Kensington, was constructed. This unit is believed to have been the first to see actual service on a steam engine.

It seems to be firmly established, then, that the centrifugal governor was invented in the second half of the year 1788. If Boulton's letter provided the initiative for the invention, it was assisted by excellent timing. The basic problems of the rotating steam engine had finally been solved after several years of labor; now James Watt and his engineers had the leisure to direct their attention to refinements of the engine. Further, a new generation of engineers was beginning to come into its own at the firm of Boulton & Watt (Boulton was 60 and Watt 52), eager to prove themselves by inventions of their own. John Southern, for example, who later invented the engine indicator, and who at that time was just 30 years old, must have designed the governor with some enthusiasm. Still another circumstance favored the invention: only a few months earlier, about May

1788, a throttling valve had been developed that was originally intended only for manual control but was perfectly suited to be employed with the governor.

Watt described his own role in the invention of the centrifugal governor as follows: "The application of the centrifugal principle was not a new invention, but it had been applied by others to the regulation of water and windmills, and other things, but Mr. Watt improved the mechanism by which it acted upon the machine and adapted it to his engines."[212] He thought his contribution lay only in the application and in the practical development of a known principle, and he chose not to take out a patent for the governor.[213] When in 1793 competitors imitated the governor, Boulton & Watt decided to remain silent. At about this time Watt, who presumably knew of Mead's and Hooper's patents, made inquiries into the history of the centrifugal pendulum. Several letters he received from business associates are extant; they report on the basis of hearsay that in connection with millstones the governor had been in use for some 20 or 30 years. Watt was right, then, in not claiming the invention of the centrifugal pendulum for himself. But the application of the centrifugal pendulum in a system of speed regulation of steam engines was a new breakthrough for which the firm of Boulton & Watt, if not James Watt himself, clearly deserves the credit.

What was James Watt's attitude to the principle of feedback? To begin with, we have to explain the state of the available sources. Watt was a conscientious and graceful letter writer, but he left neither diaries nor memoirs. The subjects upon which he expressed himself in writing depended on the external occasion. In the course of the present study it was not possible to search through the mass of his correspondence and miscellaneous papers preserved in several British archives.[214] Instead we have to content ourselves with the published literature according to which Watt has written but little on the centrifugal governor and nothing on the underlying principle which we now call feedback and which he probably did not recognize in its universality. Still, he may well have sensed that the governor had something in common with other feedback mechanisms that he knew. The governor did not strike him as new; he considered it merely an adaptation of a known invention to a new task. In spite of an experienced engineer's distaste for unnecessary complications and his own cautious and skeptical temperament, Watt employed regulators for speed, water level, and steam pressure as soon as they proved advantageous in practice. One might infer from his silence that he did not see anything particularly significant in this principle. Compared with the large but straightforward task of producing power, the regulating devices and whatever theory they involved may have appeared to the sober Watt as secondary, if not marginal. One and a half centuries later, when feedback came to be regarded as a key concept not only in industrial but also in sociological matters, the character of technology had changed far beyond anything Watt could have imagined.

2. The Spread of the Centrifugal Governor

Quantitative data about the production of the governor are not available; we can estimate the spreading of the invention only from the resonance it produced in the literature.

No communications in scientific or technical journals have been found that announce the arrival of the newly invented governor. Instead of patenting it, Boulton & Watt tried to protect the device by secrecy. Only those who saw it themselves in action on a Watt engine or who heard of it through eyewitnesses learned about the governor. Nevertheless the Soho Works registered a lively demand for it. Already in late 1789 Peter Drinkwater, a customer of Boulton & Watt's, asked in a letter for delivery of a regulator; he promised to use it in a concealed place, as requested. Toward the end of 1790 John Rennie in London, asking for four or five units, wrote that "... several of our engine proprietors are very anxious about governors."[215]

Despite their policy of secrecy, Boulton & Watt could not prevent that the governor was copied by a competitor in 1793; they did manage, however, to keep it hidden in 1791 from the inquisitive Georg von Reichenbach. In his diary (see above, p. 80), the young Bavarian made no reference to the invention, although he surely would have known how to appreciate it.

Even though the governor spread only gradually, it was universally known in England by 1804. When an applicant for a patent covering the "Clothing and Unclothing of Windmills while in Motion" attempted to describe the centrifugal pendulum as applied to the speed regulation of mills, instead of giving a detailed explanation, he simply referred to it as "a pair of centrifugal balls—like the governor of the steam engine ...".[216] The development had come full circle: the centrifugal pendulum, after attaining success and fame in its association with the steam engine, returned to its starting point, the windmill.

Soon after the turn of the 19th century the governor entered the textbooks of mechanics and engineering. Thomas Young, in his *Lectures on Natural Philosophy and the Mechanical Arts* of 1807, described three applications of the centrifugal pendulum: the steam-engine governor, a regulator for chronographs, and the lift-tenter.[217] In 1811, a professor of the Paris *École polytechnique,* J. N. P. Hachette, included in his *Traité élémentaire des machines* a detailed illustrated description of the governor.[218] He did not name his source, nor did he have a proper term for the device, paraphrasing it awkwardly as a "mécanisme par lequel on règle les ouvertures des soupapes dans les Machines à feu." His source probably was not literary but a drawing or a real engine.

Buchanan's book *On Millwork and other Machinery* (1814) presented a full chapter on governing devices, which includes, apart from the classic steam-engine governor, two types of lift-tenters and five different designs for waterwheel governors.[219] This chapter was reprinted, in abridged form, and with acknowledgement of its source, in 1815 in Olinthus Gregory's *Treatise of Mechanics*.[220]

Borgnis' *Traité complet de mécanique* of 1818 treated the governor in a fashion similar to Hachette's, but he referred to it as *pendule conique de Watt*. Borgnis reports that this regulator has been used with success on several steam engines and offers his own suggestion to employ it on water-pressure motors.[221] A description of the steam-engine governor, written by James Watt himself, is added as an appendix to John Robison's *Steam and Steam Engines* of 1818.[212]

What role did the governor play in the work of Oliver Evans, the pioneer of American engineering? The French edition (1821) of his *Abortion of the Young Steam Engineer's Guide* treats the governor (termed *modérateur*) at length,[222] but this section was added by the translator; the original edition of 1805 does not mention the device.[223] However, his notebook *Directions for Putting up and Working a steam engine* of 1817 does include a description of the governor.[224] References to other feedback devices have not been found in Evans' work.

By the 1820s, the governor had gained a firm place in the engineering textbooks. It is treated extensively, for example, in the works of Christoph Bernoulli (1824),[225] Karl Christian von Langsdorf (1826),[226] J.-V. Poncelet (1826),[227] John Farey (1827),[228] Thomas Tredgold (1827),[229] or Zachariah Allen (1829).[230] Poncelet, Langsdorf, and Tredgold even show the rudiments of a quantitative treatment, although still limited to stationary behavior of the system. The problems of control dynamics had to wait until the second half of the century.[231]

The fact that the steam engine, and with it the governor, received due attention in the professional literature of the engineers does not seem surprising. More remarkable, however, is the appearance of a large number of historical works on the steam engine. This tends to indicate that the general public was eager to acquire a deeper understanding of the celebrated engine by means of the historical method. Works of this kind are, for example, the books of Partington (1822),[232] Stuart (1824),[233]

Galloway (1828),[234] as well as Arago's extensive article "Notice historique sur les machines à vapeur" of 1829, which was reprinted several times.[235] Every one of these includes detailed descriptions of the centrifugal governor.

3. The Regulator of the Brothers Périer

Initially the centrifugal governor had had a rival which by now has been entirely forgotten. The brothers Périer, who are famous for founding the first French steam-engine factory and for introducing James Watt's engine to France,[236] around 1790 designed a speed regulator based on the float valve. In the contest with the centrifugal governor this device was defeated and soon forgotten; the information available about it is therefore sparse. The earliest and most informative witness, M. R. de Prony, gives the following account in his *Nouvelle architecture hydraulique* of 1790/96:[237] In 1788, a Spanish nobleman, Augustin de Béthencourt (1760-1826) visited London and inspected the latest Watt engines (presumably at the Albion Mill), whereupon he had a model of these built in Paris. Since he had not been allowed to study the interior of the engine, he was forced to reinvent a number of components; nevertheless, the pilot engine ran most successfully. Following this model, the brothers Périer built a large double-acting rotating steam engine to be installed on the *Isle des Cygnes* in Paris, in order to provide power for a grist mill. Concerning the date of its completion de Prony is not clear: in his first volume (published 1790) he writes that the engine was intended to be put into operation by the end of 1790; the second volume of 1796 states that in the meanwhile it has been finished, without however giving the date.

De Prony referred to the speed regulator of the engine of *Isle des Cygnes* as a "modérateur" or as a "mécanisme au moyen duquel la machine conserve spontanément un mouvement uniforme et constant sans le secours d'aucun agent extérieur."[238] Figure 68 shows the lower half of the steam cylinder *3*, the condenser *R*, into which cooling water from the tank *6* is injected through the pipe *3-4-5* at *5*. The vessel *16* with the float is the heart of the regulator. Not shown are the piston, the walking beam, and the large flywheel located to the right.

The controlled variable, i.e., the engine speed, is sensed by the pump *r-r*, which is driven from above by the walking beam, injecting with each engine stroke a certain quantity of water into the vessel *16*. The float *aa* serves two purposes: first, it supports the syphon *f-g-h* through which water runs back into the tank *6* at a constant rate adjustable by the valve *k* (the syphon is *not* suspended from ropes, as Fig. 68 deceptively suggests). Second, the float is the feedback element: by means of the lever *b-c-d*

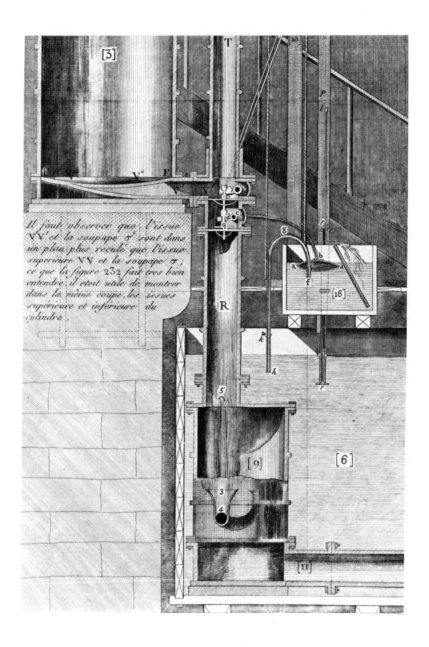

Figure 68 Speed regulation of Périer steam engine.

pivoted at c it is connected with the throttle valve q that controls the flow of the exhaust steam to the condenser.

The flow rate of the water pumped into the vessel *16* is proportional to the actual engine speed, while the desired speed is represented by the flow rate through the syphon *f-g-h* which is constant, since it occurs under a constant head. If the engine runs faster than desired, the water level in *16* will rise, causing the valve q to be closed, thus reducing the speed. The water level will stop to rise only when inflow and outflow, i.e., actual and desired speed, are equal. Expressed mathematically, the height of the water level h in the vessel *16* is related to the rate of inflow q_{in} and outflow q_{out} [volume/time], as follows:

$$q_{in} - q_{out} = \frac{dh}{dt} \times A,$$

or

$$h = \frac{1}{A} \int (q_{in} - q_{out}) \, dt + C,$$

where A is the cross-sectional area of the vessel *16*, and the constant of the integration C corresponds to an average water level at design speed. This means that as long as the error signal $q_{in} - q_{out}$ deviates from zero, the water level will continue to move in the direction that will eliminate this deviation. We are thus dealing for the first time with an *integral* control system; all previous systems provided *proportional* control characterized (retaining the above symbols) by a relationship of the form $h = K(q_{in} - q_{out}) + C$, where K is the proportional gain or sensitivity. Proportional control has the disadvantage that in case of a sustained disturbance it will tolerate a certain permanent control error, while integral control will continue its corrective action until all error is eliminated. This virtue of integral control, however, has to be compensated for by its poorer dynamic properties.

The Périers' speed regulator was of course not deliberately designed as an integral control system. This feature was rather an unintended consequence of the use of the float valve, and it was probably of no great interest to the inventors, since the desired speed was easily adjusted by means of the valve k.

The float valve speed regulator has been mentioned by several later authors. Borgnis (1818) quotes the relevant passages from Prony literally without adding anything and without acknowledging his source.[239] Partington (1822) describes a modified version of the device: the float no longer acted upon the exhaust valve but, like Watt's centrifugal governor, upon the steam inlet valve of the engine.[240] He provides no historical data, but his source was presumably also Prony's *Nouvelle architecture hydraulique,* which is cited in his bibliography. Tredgold (1827) mentions the device in a short footnote, charging it with low sensitivity and a tendency to break down.[241] Poncelet[242] analyzed the system thoroughly; he pointed out three shortcomings, loss of energy, unreliability, and sluggishness, and concluded that the centrifugal governor was decidedly superior for practical purposes. His judgment was confirmed by subsequent developments: a number of variations of the basic scheme were tried, all of which were classified by W. Trinks in 1919 as *discarded* methods of speed control.[243]

XI. The *Pendule Sympathique* of Abraham-Louis Breguet

The boldest and most imaginative system of feedback control to be dealt with in this study was applied to the synchronization of watches, a special case of speed regulation. It was invented by Abraham-Louis Breguet (1747-1823), a French-Swiss who from his 15th year lived in Paris, acquiring the reputation of the foremost watchmaker of his era. His watches, distinguished by accuracy, elegance, and a mechanical refinement unsurpassed even today, found their buyers among the highest ranks of international society.

On June 26, 1795, in a letter to his son Antoine-Louis, Breguet mentioned an invention that he considered uncommonly important:[244]

> Je te fais part avec bien du plaisir, mon ami, que j'ai imaginé une chose des plus importantes, mais sur laquelle il faut être discret, meme sur la pensée de la chose ... J'ai imaginé le moyen de mettre chaque jour une montre à l'heure et à toucher à "l'avance" et "au retard" sans que le particulier s'en occupe. La montre la plus grossièrement faite, pourvu qu'on n'arrête point, contentera celui qui la porte comme s'il avait un garde-temps. Voici comment: Il suffit d'avoir chez soi une pendule à seconde ou une montre marine disposée comme il convient pour recevoir la montre et cette dépense de surcroît est de la journée d'un ouvrier; même dépense pour la montre. Ainsi tous les soirs, en vous couchant, vous posez votre montre sur la pendule. Le matin ou une heure après, elle sera à l'heure exacte de la pendule. Il ne faut point ouvir la montre pour cela; il n'y a aucune marque extérieure qui annonce par où on y touche ... J'en conçois le plus grand avantage pour notre gloire et pour notre fortune.

It is with great pleasure that I am letting you know, my dear, that I have dreamed up a most important matter, about which you must be discreet, even when it comes to thinking about it .... I have thought of a way of setting a watch every day and of adjusting the "ahead" or "behind" lever without the owner's doing anything about it. Even the most crudely made watch, provided one doesn't let it come to a stop, will satisfy its wearer as if it were a precision instrument. Here's how it works: It is only necessary to have at home a one-second pendulum clock or a chronometer, so placed that it can accommodate the watch, the cost of which is no more than a day's wages of a common laborer plus the cost of the watch itself. Now every evening, before going to bed, you put your watch on the clock. The next morning, or even an hour later, it will be set at exactly the same hour as the clock. It is not at all necessary to open the watch in order to do this, and there is no outside mark to show where it has to be touched.... I foresee great things from it, for our fame and our fortune.

The invention, which H. von Bertele believes to date from about 1793, was named *pendule sympathique.* Three specimens, built in the beginning of the 19th century, are extant and operational to this day. Unfortunately, primary sources on Breguet's work are inaccessible and secondary literature sparse. The material presented here we owe mostly to an article by H. von Bertele.[245]

To describe the character of the invention, we must begin with a few remarks about ordinary pocket watches. Mechanical clocks employ a repeated cyclical movement, the rotary oscillation of the escapement, as a time standard. The remarkable uniformity of this movement is not the result of feedback control; it is achieved by simply protecting the oscillatory mechanism from, or by compensating it for, external disturbances (such as wear, dirt, and fluctuations in temperature, air pressure, or humidity) by means of careful mechanical design. The speed of the watch can be varied slightly by shifting a regulating arm accessible from outside which determines the frequency of oscillation by changing the initial compression of the balance spring. The usual method of regulating a watch consists of finding the correct position of the regulating arm empirically by trial and error and by comparison with a standard chronometer of known accuracy.

In Breguet's *pendule sympathique* this procedure is carried out automatically with the help of feedback. The system consists of two clocks (Fig. 69). A large, accurate table chronometer of conventional design has only one unusual feature: its top is shaped like a saddle into which the second watch can be placed; and here, punctually at twelve o'clock, a small pin emerges from the inside of the chronometer. It is the second watch that plays the main part in this system: it is a pocket watch equipped with a complicated regulating mechanism. In order to syn-

Figure 69 A.-L. Breguet's *Pendule Sympathique*.

chronize the pocket watch, one places it upon the chronometer once or twice a day, slightly before twelve o'clock, at which time the emerging pin initiates the process of regulation. Breguet had visualized that the watch was to be placed upon the chronometer every single night, but experience has shown that a very good state of synchronization is achieved in as little as two or three days.

As mentioned, the regulating mechanism itself is located in the pocket watch, the smaller of the two clockworks. The larger table chronometer has to do no more than to project the pin punctually at the stroke of twelve, a relatively simple task which any cuckoo clock can fulfill. The regulating mechanism of the pocket watch can be studied on an original drawing by Breguet (Fig. 70). This sketch, however, is not complete; it would have to be supplemented as follows: the two large four-spoke wheels are gears engaging into each other; attached to the front of the rim

Figure 70 Breguet's automatic synchronizing mechanism.

of each wheel, at the end of each spoke (barely visible on the right wheel) there is one metal lug; the segment-shaped (light) design element in the middle of the figure also has teeth on its upper perimeter.

The lower part of the figure shows the balance wheel and hair spring, and, extending upward from the center of these, the regulating arm which is linked to the lever with the segment-shaped upper part. The circular component (top, center) is attached to a lever (pointing straight down) with a (dark) arrow-shaped head (termed *minute-hand arm*): this device is coupled to the minute hand of the watch, and it rotates with it once every hour. Upon the full hour the minute-hand arm points vertically downward. If the watch is placed upon the chronometer, the pin mentioned above is inserted into the watch through an especially provided opening at the instant of twelve o'clock. Here it triggers a mechanism that makes the two large gears rotate in opposite directions, the left one clockwise, the right one counterclockwise. If the movement of the watch is not synchrononous with that of the chronometer, the angular position of the arrow-shaped minute-hand arm will deviate from the vertical by a small angle either to the right or left. Now the lugs on the large gears and the peculiar-shaped levers on both sides of the arrow-head will come into play: the longer pair of levers (shaded) are *feelers,* the shorter ones (white) are *pawls.* The two pairs are located in different planes. The feelers lie in the plane of the lugs, whereas the pawls face the fine teeth of the segment-shaped lever, although without engaging it yet. When precisely at twelve o'clock the lugs on the two large gears begin to move in from both sides toward the minute-hand arm, and if the latter is not pointing straight down, one of the lugs will hit the left or right feeler (depending on whether the watch is fast or slow), producing two effects: first, it pushes the minute-hand arm into the correct vertical position, thus setting the watch; second, it sets into motion a special mechanism which presses the pawls outward until they engage the teeth of the gear segment. The gear segment is rotated by a small angle, thereby bringing the regulating arm into a new position. This results in a more accurately running watch. If the pocket watch runs in exact synchronization with the chronometer, the minute-hand arm will point straight down at the instant when the chronometer indicates 12 o'clock, and therefore it will be bypassed by the revolving lugs.

Since the desired (synchronous) speed of the watch is compared with the measured actual speed, and since the movement of the watch is either decelerated or accelerated on the basis of the result of this comparison, the *pendule sympathique* represents true feedback control. This is confirmed by the block diagram of Fig. 71. The system displays however a few peculiarities: In ordinary feedback control systems the command variable is adjustable, to be set according to the requirements of the particular task. In the case of the speed regulation of a watch the only conceivable value of the command variable can be synchronism with some ideally perfect chronometer. Therefore there can be no motive for varying the command signal. Second, it is evident that the system works discontinuously. According to the criteria adopted at the outset this is permissible, for continuity of operation is not an essential property of feedback control. In fact, discontinuous control systems such as the *pendule sympathique* are not uncommon today, and are referred to as *sampled-data control systems.*

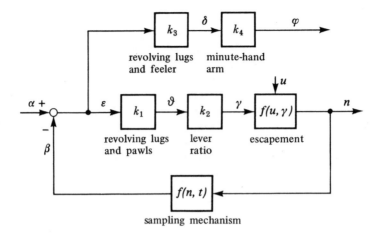

a = desired synchronous speed
(desired position of minute-hand arm perpendicularly down)
β = actual position of minute-hand arm
ϵ = $a - \beta$
δ = angular correction of minute-hand arm
φ = angular position of minute-hand arm
ϑ = angular correction of segment lever
γ = angular correction on regulating lever of escapement
n = speed of watch
u = disturbances

Figure 71 Breguet's automatic synchronizing mechanism, block diagram.

The *pendule sympathique* does not seem to be historically connected with the earlier inventions representing feedback control. It is not known what has inspired Breguet to his ingenious use of the idea of the closed loop. He may have been familiar with Bonnemain's thermostat, or with the steam-engine regulators of Watt or the Périers. If so, then he owed to such models no more than the concept of feedback in its most abstract form.

XII. Concluding Remarks

The results of this study are presented graphically in a quasigenealogical chart (Fig. 72). Along a time axis (the time scale is expanded after A.D. 1600) there are marked only inventions of genuine feedback devices, each labeled with the name of the inventor. Borderline cases and repetitions not claiming originality are not included. This chart gives rise to a few questions.

1. Completeness

Since the source material available for the period prior to the nineteenth century is manageable in size, an attempt was made to present exhaustively all inventions of feedback devices as defined. Such completeness, however, was not attainable. Three gaps that could not be filled in spite of considerable efforts are indicated by question marks.

(a) For float regulators in Europe, the earliest reference is dated 1746 (it is found in the third edition of W. Salmon's book;[150] the dates of earlier editions could not be established). Soon thereafter, the float regulator appeared in Brindley's patent for the regulation of boiler feed. It is unlikely that Brindley was inspired by Salmon's book.[150] More probably both are based on an older model which remains unknown to us.

(b) A similar problem exists in the area of millwrighting. The first patent involving the centrifugal regulator was granted to Mead in 1787; Watt's governor was invented in 1788, and Hooper's patent followed in 1789. But

126 CONCLUSIONS

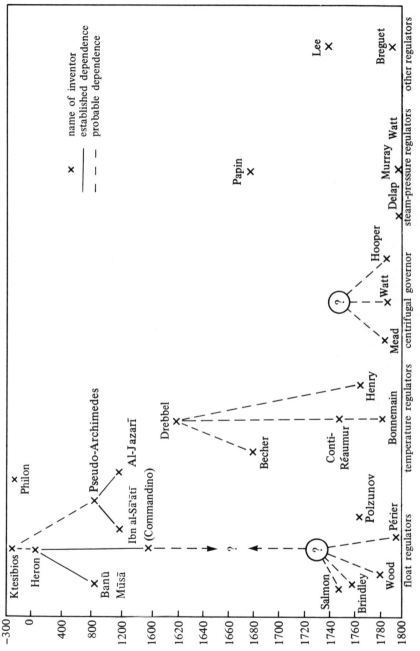

Figure 72 Evolution of feedback devices, chronological diagram.

a number of witnesses maintain that such devices had been used by millwrights years earlier. Although few subjects in the history of technology have been studied as extensively as the steam engine and windmills in 18th century Britain, no evidence has been found to support these claims.

(c) While these two cases involve merely gaps in our own knowledge, there is another instance of discontinuity which seems to exist in the historical events themselves.

The tradition of the ancient float regulator ends with two Arabic horological manuals written around A.D. 1200. No such float regulators could be discovered in the European literature of the Middle Ages or the Renaissance, still less regulators of other types. More astonishing yet is the following: Commandino's translation of Heron's *Pneumatica* of 1575 contains several descriptions of float regulators, accompanied by clear illustrations. But the many authors of the Baroque period who have reproduced Heron's other pneumatic apparatus with obvious delight, ignored the float regulator. Their attitude has all the appearance of unanimous, deliberate rejection. This gap in the history of the float regulator extends from the early 13th to the middle of the 18th century; we have to assume that it exists not in our historical knowledge but in the events themselves.

2. The Rise during the 18th Century

Admittedly, the listing of inventions presented on Figure 72 is incomplete, the inventions listed may differ in importance, and perhaps the criteria by which we have classified inventions as feedback devices, or as being independent, may not be entirely without arbitrariness. Yet, regardless of such reservations this form of presentation leads to some intriguing considerations.

Drebbel's thermostatic furnace of ca. 1620 is the first feedback system invented since antiquity. If in addition we count also the proposal of Becher (1680) and Papin's safety valve (1681), then we know of only three relevant inventions in Europe prior to 1745, in contrast to no less than fourteen between 1745 and 1800. In the middle of the 18th century, then, the application of feedback in technology achieves a veritable breakthrough. What accounts for this striking phenomenon?

Purely technological factors were probably not decisive; almost all the 18th-century regulating devices could have been built with the practical know-how and the scientific accomplishments of the Renaissance.

The argument that progress in other fields of technology had created a need of regulating devices and had thus initiated the lively inventive activity must be taken more seriously. As examples, one might point to the progress in domestic water supplies, which necessitated float regulators, or to the steam engine. Half of the inventions listed in Figure 72 originate from this latter area (the inventions of Brindley 1758, Polzunov 1765, Wood 1784, Watt 1788, Périer ca. 1790, Delap 1799, Murray 1799). These various steam-engine regulators soon found widespread industrial employment, while the other seven inventions remained relatively little known. It does not follow from this, however, that the steam engine was the sole cause of the increasing interest in feedback regulation. On the contrary, feedback came to the steam engine relatively late (from 1712 up to 1758 the Newcomen engine had to manage without the float regulator), while important feedback devices had already been applied in other fields. Some feedback regulators designed for uses other than the steam engine were very slow in finding industrial application; this contradicts the claim that they were invented to satisfy a definite demand.

In the same country and at the same time as the breakthrough of feedback control there took place a historical process of incomparably greater significance, the Industrial Revolution, which manifested itself in a flood of inventions and innovations in such diverse fields as agriculture, transportation, manufacture, commerce, and finance. The phenomenon with which we are concerned is undoubtedly also a symptom of the Industrial Revolution, but classifying it thus does not explain it.

Neither purely technical factors nor events in the economic environment have been able to account for the increased use of the concept of feedback in the 18th century. Since the inventions involved are most varied in external form, and since their inventors for the greater part worked independently of each other, one suspects that the true causes for this upswing are to be found in the intellectual currents of the era. Between the 17th and the 18th centuries the attitudes toward the problem of regulation underwent a radical change. The technicians of the Renaissance and Baroque loved rigidly programmed control, as symbolized by automatons, while they disregarded the idea of feedback in an almost conspicuous manner. By the second half of the 18th century, however, the more progressive engineers in Britain and France had learned to appreciate the concept and did not hesitate to employ it in a variety of fields.

The analogous development took place in the field of political economy. Mercantilism, especially of Colbert's stamp, taught that for a nation the path to prosperity was a carefully planned economic policy, centrally directed by statesmen of great competence.

The economic system of liberalism, formulated comprehensively in 1776 by Adam Smith in his *Wealth of Nations,* in contrast, was based upon the postulate of *laissez faire:* if only the state would refrain from all interference, the organism of the economy would automatically swing into equilibrium at optimal conditions. Instead of the central control, the driving force here is the self-interest of the individual participants in economic life. Smith's theory that deviations from an optimal state of the economy would automatically be corrected by the system of free enterprise and the law of supply and demand implies a conception of closed causal loop that is in principle the same as the feedback loop.[246]

Adam Smith's book was a brilliant summary of the system of economic liberalism that had evolved during his time, but he was not the sole originator of all its intellectual content. He had a forerunner in the concept of economic systems as self-regulating mechanisms: David Hume in his essay *On the Balance of Trade* (1752) had formulated such a theory in connection with the mechanics of international specie distribution. Arguing in favor of free trade and against the doctrine of a positive balance of trade, he asserted that, given enough time, the balance of trade of any country would automatically swing into equilibrium; for if a country had a surplus of money and a shortage of goods, it would necessarily begin to trade with some country that lacked money.[247]

The similarity between the two antithetical pairs *Mercantilism–Liberalism* and *automaton–feedback control* does not, of course, represent evidence of historical connections between the two areas, and indeed it would be wrong to interpret one of these developments as a direct consequence of the other. A theoretician like Adam Smith may have observed feedback devices in operation, and this may have made his analysis sharper and his formulations more concrete, but the beginnings of economic liberalism lie in the early 18th century, antedating the breakthrough of the feedback concept in technology. On the other hand it is also highly unlikely that 18th-century inventors should have obtained their conceptions by materializing abstract theories of political economy.

3. The General Concept of the Closed Feedback Loop

The abstract concept of the closed causal loop which provides the common basis for all the regulating mechanisms discussed in this study, and which is expressed most compellingly by the graphic symbols of the block diagram, is an achievement of the 20th century. As we observe how the inventors of the 18th century gradually become more familiar with the practical applications of feedback, the question arises how far they progressed toward a general and abstract formulation of the concept.

At the end of the 18th century feedback mechanisms in English were referred to as *regulators* or *governors,* while the French employed the terms *régulateur* and *modérateur:* (Boyle, 1660, "a furnace which he could govern . . ." (p. 56); Dr. Goddard, 1662, "Drebbel's method of governing a furnace . . ." (p. 57); Rennie, 1790, "governor" (p. 113); Lee, 1745, "self-regulating wind machine" (p. 93); Bonnemain, 1882, "régulateur du feu" (p. 71); Wood, 1784, "regulator" (p. 78); Mead, 1787, "regulator on a new principle" (p. 101); Watt, 1788, "centrifugal speed regulator" (p. 111); Prony, 1796, "modérateur" (p. 115). But all these terms have also been applied to regulating devices without feedback (e.g. *regulator* for clock pendulums, or, in the case of Prony, for steam-engine valve gear), while actual feedback devices occasionally had to settle for wordy paraphrases (e.g., Prony, see p. 115).

When the term *regulator* is applied to feedback devices of diverse types (centrifugal governors, thermostats, fan-tails), one might be tempted to conclude that this expressed an awareness of the one underlying principle that all these devices have in common. Such a conclusion, however, will prove false as soon as one looks up the word *regulator* in representative encyclopedias of the period. In 1812, the large *Ökonomisch-technologische Enzyklopadie* of J. G. Krünitz defines the word *Regulator* (the words *moderator* and *governor* or *gouverneur* are not listed) as any device that will make machines run more smoothly, especially in the event of irregularities in the driving force.[248] Among the examples listed are the flywheel, the balance spring and pendulum of clocks, Bonnemain's temperature regulator, and a certain tool for adjusting roller-presses used to produce sheet-lead. In Rees's *Cyclopaedia* (vol. 29, 1819) the listing *regulator* gives an extensive description of several types of steam-engine governor and other feedback devices, but it also includes a variety of devices without feedback.[249] The 6th edition of the *Encyclopaedia Britannica* (1823) lists under *regulator* only the balance spring of clocks (it has no entries for *governor* and *moderator* in their technological sense, nor is the governor mentioned in the section on the steam engine).[250] The *Dictionnaire technologique* (vol. 18, 1831) is more felicitous in its choice of examples: under the heading *régulateur* it lists, beside the clock pendulum, Watt's governor and the thermostat of Bonnemain.[251]

The term *régulateur* plays an important role in some early 19th-century French textbooks on mechanical engineering which are characterized by an attempt to present their material in a logical system of classification. In Borgnis's book (1818)[252] the main category *régulateur* is subdivided into *modérateurs, directeurs,* and *correcteurs* which in turn are reduced to three different lower ranks (*genre, espèce* and *variété*). This provides room, side by side, for flywheels, escapements, free-wheeling clutches, ratchets, brakes, etc. The centrifugal governor of Watt, the only feedback device among all these, is presented here as a special case of a *limitateur,* which in turn belongs to the group of *directeurs*. Our governor then is subordinated under the concept *regulateur* as a speciality of the fourth rank. The meaning of the term *regulator* is here so broad that devices incorporating feedback disappear in this large category. This is confirmed by the definition given by Borgnis in his *Dictionnaire de mécanique* (1823):[253]

> Régulateur, s.m.: Nom générique que l'on donne aux organes qui ont pour but de régler le mouvement des machines, et de corriger les irrégularités de leur mouvement.

> Régulateur, n.: Generic term given to devices whose purpose it is to regulate the movements of machines and to correct any irregularities in the motions.

In short, any device whose function conforms with the literal meaning of the word, regardless of the internal mode of operation, can be termed *regulator*. In the time span covered in this study, and indeed far beyond it, no author has been discovered who has referred to the closed-loop mode of action as the common principle underlying a definite class of control devices. The inventors who did design feedback mechanisms have employed the principle intuitively.

4. Subsequent Developments[254]

During the 19th century countless feedback devices were invented for a multitude of purposes, but only the speed regulator of prime movers, the descendant of James Watt's governor, found universal acceptance. Up to the beginning of the 20th century, therefore, the technology of automatic control remained a specialty of mechanical engineering. The dynamic problems of speed regulation provided the motive for the first attempts at formulating a mathematical theory of automatic control (G. B. Airy 1840/51,[255] J. C. Maxwell 1868,[256] I. I. Vyshnegradskii 1876,[257] E. J. Routh 1877,[258] A. M. Lyapunov 1892,[259] A. Stodola 1893/94,[260] A. Hurwitz 1895[261]). These attempts consisted essentially in solving the differential equations by classical methods; this was difficult and did not favor the evolution of generalized methods and concepts. Practical engineers, if they employed any theory at all, therefore preferred graphical methods, such as that of S. M. Tolle (1895).[262] By the turn of the century the classical theory of speed regulation was fully developed, presented in numerous textbooks and handbooks, and even became the subject of a historical study.[231]

The predominance of mechanical methods, both theoretical and practical, over control engineering came to an end with the rise of electrical technology. New electrical solutions were proposed for traditional control

CONCLUSIONS

problems, such as electrical speed, level, and temperature regulators; the principle of feedback proved particularly useful in the technologies of communication: the unfamiliar problems arising here required and favored theoretical treatment. Laplace-transform and operational calculus furnished the base for a new theory of automatic control, which found its first expression in the pioneering papers of K. Küpfmüller (1928),[263] H. Nyquist (1932),[264] H. L. Hazen[265] and H. S. Black (1934),[266] and which was first presented comprehensively in the books of R. C. Oldenbourg and H. Sartorius (1944),[267] and of LeRoy MacColl (1945).[268]

Practical control engineering made great progress during the Second World War, when each belligerent made efforts to gain superiority in this field. When after the war the secrecy was lifted, there suddenly became available (1) a mature technology of automatic control which had proven itself in dealing with the problems of radar, fire control, autopilots, guided missiles, and so on ; (2) a theory that was universal and easy to manipulate; and (3) a staff of scientists and engineers who quickly spread this new knowledge, thus introducing the era of automation and cybernetics.

Notes

1. The only scholarly studies on the history of feedback control known to me are the following:
   Wilhelm Hort, "Die Entwicklung des Problems der stetigen Kraftmaschinenregelung," *Zeitschrift für Mathematik und Physik* 50(1904): 233-279.
   A. R. J. Ramsey, "The Thermostat or Heat Governor: An Outline of its History," *Transactions of the Newcomen Society* 25(1945-47): 53-72.
   H. G. Conway, "Some Notes on the Origins of Mechanical Servo Mechanisms," ibid. 29(1953-55):55-75.
   A. V. Khramoi, *The History of Automation in Russia before 1917*, trans. Israel Program for Scientific Translations (Jerusalem, 1969).
   Aleksandr W. Chramoi, "Überblick über die Entwicklung der selbsttätigen Regelung in der UdSSR." In W. W. Solodownikow, *Grundlagen der selbsttätigen Regelung,* trans. from Russian (Munich, 1959), Vol. 1, pp. 12-27.
   Klaus Rörentrop, *Zur Entwicklung der Regelungstechnik,* Dissertation, University of Erlangen-Nürnberg (Erlangen, 1969).
2. Among the inventions described erroneously as first feedback devices are the south-pointing chariot of ancient China (see p. 49), the mill-hopper (*baille-blé*) (p. 90), the fly-wheel with centrifugal weights (p. 102), and C. Huygens' centrifugal pendulum (p. 102).
3. André-Marie Ampère, *Essai sur la philosophie des sciences,* Part 2 (Paris, 1843), pp. 140-41.
4. "We have decided to call the entire field of control and communication theory, whether in the machine or in the animal, by the name *Cybernetics*, which we form from the Greek χυβερνήτης or *steersman*. In choosing this term, we wish to recognize that the first signifi-

cant paper on feedback mechanisms is an article on governors, which was published by Clark Maxwell in 1868, and that *governor* is derived from the Latin corruption of χυβερνήτης." Norbert Wiener, *Cybernetics, or Control and Communication in the Animal and the Machine* (Cambridge, Mass., 1948), pp. 11-12.

5   For a detailed exposition of the block diagram notation see, for example, John J. D'Azzo and Constantine H. Houpis, *Feedback Control System Analysis and Synthesis* (New York, 1960), chapter 5 and appendix C.

6   H. W. Dickinson and Rhys Jenkins, *James Watt and the Steam Engine* (Oxford, 1927), plate 81.

7   A. I. E. E. Committee Report, "Proposed Symbols and Terms for Feedback Control Systems", *Electrical Engineering* 70(1951): 905–909, especially page 909.

8   British Standard Institution, *Glossary of Terms Used in Automatic Controlling and Regulating Systems*. B.S. 1523:Part 1:1967 (London, 1967), items no. 2002, 3202, and 3204, respectively.

9   Norbert Wiener, *The Human Use of Human Beings: Cybernetics and Society* (Garden City, N.Y., 1954), p. 61.

10 The British Standard Institution defines *self-regulation* as "The property of a body, process, or machines (without closed loop control) of reaching a new steady state after a sustained disturbance." B.S. 1523:Part 1:1967, (London, 1967), item no. 6021.

11 Transfer functions in general can be brought into the standard form

$$G(s) = \frac{K_n(1+a_1s+a_2s^2+\cdots+a_ws^w)}{s^n(1+b_1s+b_2s^2+\cdots+b_us^u)}, \tag{1}$$

(D'Azzo and Houpis, *Feedback Control System Analysis*, p. 113), which for $n = 0$ reduces to

$$G(s) = \frac{K_0(1+a_1s+a_2s^2+\cdots+a_ws^w)}{1+b_1s+b_2s^2+\cdots+b_us^u}. \tag{2}$$

The transfer function of closed feedback loops, however, is (ibid., p. 97)

$$\frac{C(s)}{R(s)} = \frac{F(s)}{1+F(s)\cdot H(s)}. \tag{3}$$

If $F(s) = K_0(1+a_1s+a_2s^2+\cdots+a_ws^w)$,

and $H(s) = \dfrac{b_1s+b_2s^2+\cdots+b_us^u}{K_0(1+a_1s+a_2s^2+\cdots+a_ws^w)}$, (4)

$$\frac{C(s)}{R(s)} = \frac{F(s)}{1+F(s)\cdot H(s)} = G(s),$$

as can easily be verified by substituting (4) into (3).

This shows that all transfer functions of the type of equation (2) can formally be represented as closed feedback loops.

12 Literature on Ktesibios:
    A. G. Drachmann, *Ktesibios, Philon, and Heron* (Copenhagen, 1948).
    K. Orinsky, "Ktesibios," *Pauly's Realencyclopädie* 11.2(1922): 2074-76.
    K. Tittel, "Hydraulis," ibid. 9.1(1914):63ff.
13 A. R. Hall assigns later dates to Ktesibios. "In order to not attribute to them excessive antiquity, we may cautiously assume that Ktesibios lived about the beginning of the first century B.C. . . ." He supplies no evidence for this apart from the tacit assumption that historical dates would be the safer the later. A. R. Hall, "Military Technology." In: C. Singer, et al., *A History of Technology* (Oxford, 1956), vol. 2, p. 708.
14 A. G. Drachmann, "On the Alleged Second Ktesibios," *Centaurus* 2(1951):1-10.
15 Drachmann, *Ktesibios, Philon, and Heron*, p. 184.
16 Vitruvius *De architectura libri decem* 1.1.7.
17 Vitruvius 9.8.4-7.
18 Translation by O. M.
19 Albert Rehm, "Neue Beiträge zur Kenntnis der antiken Wasseruhren," *Sitzungsberichte der Bayerischen Akademie der Wissenschaften*, 1920:Phil.-hist. Klasse Nr. 17 (Munich, 1921), p. 23.
20 Ibid., pp. 21-23.
21 Drachmann, *Ktesibios, Philon, and Heron*, pp. 17-21.
22 Ibid., p. 20.
23 Hermann Diels, *Antike Technik*, 3rd ed. (Leipzig, 1924), pp. 203-207.
24 Rehm, *Zur Kenntnis der antiken Wasseruhren*, pp. 15-19.
25 Drachmann, *Ktesibios, Philon, and Heron*, p. 34.
26 Ibid., p. 35.
27 Literature on Philon:
    Drachmann, *Ktesibios, Philon, and Heron*.
    K. Orinsky, O. Neugebauer, A. G. Drachmann, "Philon von Byzanz," *Pauly's Realencyclopädie* 20.1(1941):53-54.
28 Philon, *Belopoiika*, Greek and German, trans. H. Diels and E. Schramm, *Abhandlungen der preussischen Akademie der Wissenschaften*, 1918:Phil-hist. Klasse Nr. 16(Berlin,1919), 51.17.
29 Philon, *Pneumatica*, Arabic version (Le livre des appareils pneumatiques et des machines hydrauliques), Arabic and French, trans. Carra de Vaux (Paris, 1902).
30 Philon, *Pneumatica*, Latin version (Liber Philonis de ingeniis spiritualibus), Latin and German, trans. W. Schmidt. In: *Heronis Alexandrini opera quae supersunt omnia*, vol. 1 (Leipzig, 1899), pp. 458-489.
31 Philon, *Pneumatica*, Arabic version, p. 117.
32 Philon, *Pneumatica*, Latin version, pp. 487-89.
33 Literature on Heron:
    Drachmann, *Ktesibios, Philon, and Heron*.
    K. Tittel, "Heron," *Pauly's Realencyclopädie* 8.1(1912):992-1080.

34 For details on Heron's dates, see O. Neugebauer, "Über eine Methode zur Distanzbestimmung Alexandria-Rom bei Heron," *Det Kgl. Danske Videnskabernes Selskab. Historisk-filologiske Meddelelser,* 26.2 (Copenhagen, 1938).
A. G. Drachmann, "Heron and Ptolemaios," *Centaurus* 1(1950): 117-131.
35 Heron, *Pneumatics* (The Pneumatics of Hero of Alexandria), trans. Bennet Woodcroft (London, 1851).
36 Heron, *Heronis Alexandrini opera quae supersunt omnia,* vol. 1. *Pneumatica et automata,* Greek and German, trans. W. Schmidt (Leipzig, 1899).
37 Wilhelm Schmidt, *Die Geschichte der Textüberlieferung,* supplement to the above, pp. 142 ff.
38 Heron, *Opera . . . omnia,* vol. 1. *Automata* 2.12 (trans. W. Schmidt, p. 349).
39 The drawings, chapter headings, and chapter numbers used here are those of Schmidt's edition. As text and drawings are in perfect agreement, I have dispensed with repeating Heron's description of the devices in question.
40 Heron, *Automata* 10.1. *Opera . . . ommia,* vol. 1 (W. Schmidt), p. 404.
41 Drachmann, *Ktesibios, Philon, and Heron* pp. 50-59 and 102-06.
42 Derek J. de Solla Price, "An Ancient Greek Computer," *Scientific American* 200 (June 1959):60-67.
43 Published from a 13th-century manuscript of the Vatican, in: Hermann Diels, "Über die von Prokop beschriebene Kunstuhr von Gaza," *Abhandlungen der preussischen Akademie der Wissenschaften,* 1917: Phil.-hist. Klasse Nr. 7 (Berlin 1917).
44 Eilhard Wiedemann, Fritz Hauser, "Uhr des Archimedes und zwei andere Vorrichtungen," *Nova Acta. Abhandlungen der Kaiserl. Leopold.-Carol. Deutschen Akademie der Naturforscher* 103:2 (Halle, 1918).
Drachmann, *Ktesibios, Philon, and Heron,* pp. 36-41.
45 Wiedemann, Hauser, "Uhr des Archimedes," pp. 164-165.
46 Ibid., pp. 172-173.
47 Ibid., p. 165.
48 Ibid., p. 169.
49 Eilhard Wiedemann, "Über die Uhren im Bereich der islamischen Kultur," *Nova Acta. Abhandlungen der Kaiserl. Leopold.-Carol. Deutschen Akademie der Naturforscher* 100:5 (Halle, 1915), pp. 42-166.
50 Ibid., pp. 60-61, 67-69, 70-72.
51 Ibid., pp. 82 and 106: compare Figures 38 and 51.
52 Ibid., p. 82.
53 Ibid., p. 66.
54 Ibid., p. 70.
55 Ibid., pp. 167-266.
56 Ibid., p. 168.
57 Ibid., pp. 168-171.
58 Ibid., p. 179.

59 Ibid., p. 191.
60 Alfonso X. de Castilla, *Libros del Saber de astronomía*, ed. Manuel Rico y Sinobas, 5 vols. (Madrid, 1863-67), vol. 4, pp. 1-108.
61 Literature on the clocks of Alfonso X:
Gustav Bilfinger, *Die mittelalterlichen Horen* (Stuttgart, 1892), pp. 151-155.
Alfred Wegener, "Die astronomischen Werke Alfons X.," *Bibliotheca Mathematica*, 3. Folge, 6 (1905): 129-185, esp. 162-164.
Franz Maria Feldhaus, "Die Uhren des Königs Alfons X. von Spanien," *Deutsche Uhrmacherzeitung* 54 (1930):608-612.
62 Silvio Bedini, "The Compartmented Cylindrical Clepsydra," *Technology and Culture* 3 (1962): 115-141.
63 Leonardo da Vinci, *Il Codice Atlantico nella Bibliotheca Ambrosiana di Milano* (Milan, 1894), fol. 20rb, 20vb, 288ra, 298va, 343a, 363va, 373vb.
64 Wiedemann, "Über die Uhren im Bereich der islamischen Kultur," pp. 11, 14.
65 Marie Boas, "Heron's Pneumatica: A Study of its Transmission and Influence," *Isis* 40 (1949): 38-48.
66 Wilhelm Schmidt, *Die Geschichte der Textüberlieferung*, supplement to vol. 1 of *Heronis:... opera...omnia* (Leipzig, 1899), pp. 142-143.
67 Ibid., pp. 67-69.
68 K. Tittel, "Heron," *Paulys Realencyclopädie* 8.1 (1912):1074.
69 Friedrich Hauser, "Über das kitāb al ḥijal—das Werk über die sinnreichen Anordnungen—der Benū Mūsā," *Abhandlungen zur Geschichte der Naturwissenschaften und der Medizin*, No. 1 (Erlangen, 1922).
E. Wiedemann, and F. Hauser, "Über Trinkgefässe und Tafelaufsätze nach al-Ġazarī und den Benū Mūsā," *Der Islam* 8 (1918):55-93, 268-291.
70 Aldo Mieli, *La Science arabe et son rôle dans l'évolution scientifique mondiale* (Leiden, 1938), p. 71.
71 De Lacy O'Leary, *How Greek Science Passed to the Arabs* (London, 1951), p.165.
72 E. Wiedemann, *Beiträge zur Geschichte der Naturwissenschaft* (Erlangen), 3 (1905):228, 230, 251; 4-5 (1905):424, 438; 6 (1906):2.
73 Hauser, "Über das kitāb al ḥijal...," pp. 32-33.
74 E. Wiedemann, ibid., 3(1905):221: 4-5(1905):424.
75 Wiedemann and Hauser, "Über Trinkgefässe und Tafelaufsätze...," pp. 270-71.
76 Ibid., p. 271.
77 Ibid., pp. 272-74, 281-82.
78 Ibid., pp. 277-79, 282-84.
79 Ibid., p. 274-77.
80 Hauser, "Über das kitāb al ḥijal...," pp. 163-67.
81 Wiedemann and Hauser, "Über Trinkgefässe und Tafelaufsätze...," pp. 279-281.
82 E. Wiedemann, *Beiträge zur Geschichte der Naturwissenschaften* 12-13(1907): 201-202.

138 NOTES

83 C. H. Haskins, *Studies in the History of Medieval Science* (Cambridge, Mass., 1924), pp. 181-83.
84 Schmidt, *Die Geschichte der Textüberlieferung*, pp. 3-6.
85 Boas, "Heron's Pneumatica," p. 40.
86 Valentin Rose, *Anecdota graeca et graecalatina* (Berlin, 1870), vol. 2 pp. 286-290.
87 Joseph Needham, *Science and Civilisation in China* (Cambridge, 1954-), vol. 3 and vol. 4.2.
88 Ibid., vol. 4.2, pp. 286-303.
89 Ibid., vol. 4.2, p. 303.
90 Ibid., vol. 3, p. 314.
91 Ibid., vol. 3, p. 317.
92 J. Needham, Wang Ling, D. J. Price, *Heavenly Clockworks* (Cambridge, 1960), p. 40.
93 Ibid., p. 32 and fig. 8.
94 Ibid., p. 91 and fig. 39.
95 F. M. Jaeger, *Cornelis Drebbel en zijne Tijdgenooten* (Groningen, 1922).
Gerrit Tierie, *Cornelis Drebbel* (Amsterdam, 1932).
F. W. Gibbs, "The Furnaces and Thermometers of Cornelis Drebbel," *Annals of Science* 6(1948):32-43.
96 Library of Carpentras, Manuscript no. 1776, fol. 409. Quoted from Jaeger, *Cornelis Drebbel*, p. 126.
97 *Nouvelle Biographie Générale,* vol. 39 (Paris, 1862), s.v. Peiresc.
98 British Patent (Old Series) No. 75, 1634: Hildebrand Prusen and Howard Stachy, "Stoves and Furnaces for the Manufacture of Salt."
99 Robert Boyle, *Works,* ed. Thomas Birch, 6 vols. (London, 1772), vol. 5, p. 644.
100 Tierie, *Cornelis Drebbel,* pp. 44-45.
101 Balthasar de Monconys, *Journal des voyages,* 3 vols. (Lyon, 1665-66), vol. 2, pp. 53-54.
102 Thomas Birch, *History of the Royal Society of London,* 4 vols. (London, 1756-57), vol. 1, p. 119.
103 Jaeger, *Cornelis Drebbel,* pp. 49-50.
104 de Monconys, *Journal,* vol. 2, p. 41.
105 Cambridge University Library, Manuscript no. 2206, L1.5.8; Catal. IV (1861), p. 63. Quoted from Jaeger, *Cornelis Drebbel,* pp. 135-138.
106 Jaeger, *Cornelis Drebbel,* p. 48.
107 Ibid., pp. 137-138.
108 de Monconys, *Journal,* vol. 2, pp. 41-42.
109 Tierie, *Cornelis Drebbel,* pp. 21-28.
110 E. J. Holmyard, "Alchemical Equipment," Charles Singer et al., *A History of Technology,* 5 vols. (London and New York, 1954-58), vol. 2, p. 743.
111 Francis Bacon probably had Drebbel in mind when he wrote: "We knew a Dutchman, that had wrought himself into the belief of a great person, by undertaking that he could make gold: whose discourse was, that gold might be made; but that the alchymists overfired the work: for (he said) the making of gold did require a very

temperate heat, as being in nature a subterrany work, where little heat cometh; but yet more to the making of gold than of any other metal; and therefore that he would do it with a great lamp, that should carry a temperate and equal heat; and that it was the work of many months."

Quoted from F. W. Gibbs, "The Furnaces and Thermometers of Cornelis Drebbel," *Annals of Science* 6(1948):36.

112 Daniel Schwenter, *Deliciae physico-mathematicae, oder Mathematische und philosophische Erquickstunden* (Nuremberg, 1636), pp. 446-447.
113 Robert Hooke, *Lampas, or: Description of some Mechanical Improvements of Lamps and Waterpoises* (London, 1677), pp. 41-42.
114 Herbert Hasssinger, *Johann Joachim Becher* (Vienna, 1951).
115 Johann Joachim Becher, *Närrische Weissheit und Weise Narrheit*, 1st ed. 1682, (n.p., 1707), pp. 86-88.
116 Johann Joachim Becher, *Minera arenaria perpetua*, 3rd supplement to *Physica subterranea*, 1st ed. 1680, (Leipzig, 1738), pp. 499-500.
117 Jean Torlais, *Réaumur* (Paris, 1961).
    Pierre-P. Grassé et al., *La vie et l'oeuvre de Réaumur* (Paris, 1962).
118 Torlais, *Réaumur*, pp. 303-314.
119 Grassé, *La vie... de Réaumur*, pp. 9-10.
120 René-Antoine Ferchault de Réaumur, *Art de faire éclorre et d'élever en toute saison des oiseaux domestiques de toutes espèces*, 2 vols., 2nd ed. (Paris, 1751), vol. 1, pp. 172-183.
121 Ibid., p. 174.
122 Grassé, *La vie... de Réaumur*, p. 23.
123 Réaumur, *Art de faire éclorre*, p. 183.
124 *Dictionnaire de Biographie Française*, vol. 9 (Paris, 1961), s.v. Conti(9).
125 *Dictionary of American Biography*, vol. 8 (New York, 1932), s.v. Henry, William.
126 "A Description of a Self-Moving or Sentinel Register, invented by William Henry, of Lancaster," *Transactions of the American Philosophical Society* 1 (1769-1771):286-289.
127 Marie-Fernande Alphandery, *Dictionnaire des Inventeurs Français* (Paris, 1962), s.v. Bonnemain.
128 *Bulletin de la Société d'encouragement pour l'industrie nationale* 27 (1828):181-182.
129 Bonnemain, *Observations sur l'art de faire éclore et d'élever la vollaille sans le secours des poules* (Paris, 1816), p. 15.
130 Ibid., p. 24-25.
131 *Bulletin des Sciences technologiques* 3(1825):173.
132 Bonnemain, *Observation sur l'art de faire éclore*, p. 24.
133 Ibid., p. 25.
134 *Bulletin de la Société d'encouragement* 23 (1824):238-242.
135 *Dictionnaire technologique*, 22 vols. (Paris, 1822-1835), vol. 11, pp. 160-169.
136 This idea was not new; it had been used in 1726 by John Harrison in a temperature-compensated clock pendulum. A. R. J. Ramsay, "The Thermostat or Heat Governor," *Transactions of the Newcomen Society* 25(1946):53-54.

137 The text in the *Bulletin de la Société d'encouragement* interprets this device as merely a thermometer. This must be incorrect, because it contradicts the accompanying drawing which is more authentic.
138 *Lichtenberg's Magazin für das Neueste aus der Physik und Naturgeschichte* 3.3 (1786): 194.
139 G. D. B. Busch, *Handbuch der Erfindungen* (Eisenach, 1821), vol. 11, s.v. Regulator.
140 Johann Georg Krünitz, *Ökonomisch-technologische Enzyclopädie*, 242 vols. (Berlin, 1773-1858), vol. 121(1812), p. 655.
141 *Dinglers Polytechnisches Journal* 16 (1825): 285-290.
142 Ibid., 29(1828):115-121.
143 *Gill's Technological Repository* 2(1828):65.
144 Andrew Ure, *A Dictionary of Arts, Manufactures, and Mines* (London, 1839), s.v. Thermostat; Heat Regulator; and Incubation, artificial.
145 *Catalogue officiel des collections du Conservatoire National des Arts et Métiers* (Paris, 1910), vol. 6, pp. 168, 262.
146 Private communication from M. Maurice Daumas, of July 28, 1966.
147 Ure, *A Dictionary of Arts*, s.v. Thermostat.
148 A. R. J. Ramsey, "The Thermostat or Heat Governor: An Outline of its History," *Transactions of the Newcomen Society* 25(1945-1947): 25(1945-1947):53-72.
149 Lawrence Wright, *Clean and Decent* (London, 1960), p. 94.
150 I owe this reference to Mr. G. B. L. Wilson, Deputy Keeper, Science Museum, London.
151 Laurence Meynell, *James Brindley: Builder of Canals* (London, 1956).
152 Samuel Smiles, *Lives of the Engineers. Early Engineering: Vermuyden–Myddelton–Perry–James Brindley* (London, 1874), p. 148 ff; revised edition (Thomas P. Hughes, ed.) Cambridge, Mass., 1966, pp.30-171.
153 British Patent (Old Series) No. 730, 1758: James Brindley, "Draining Lands, etc."
154 Ibid., p. 2.
155 Literature on Polzunov:
A. N. Voyeykov, "I. I. Polzunov," *Russkaya Starina* 40(1883): 407-416.
Nik. von Klobukow, "Zur Geschichte der Dampfmaschine," *Prometheus* 3(1892): 810-12, 827-29.
V. V. Danilevskii, *I. I. Polzunov* (Moscow, 1940).
156 Danilevskii, *Polzunov*, Fig. 73.
157 Evgenii Pavlovich Popov, *The Dynamics of Automatic Control* (London, 1962), Fig. 1.12.
158 British Patent (Old Series) No. 1447, 1784: Sutton Thomas Wood, "Steam Engine, Boilers, etc."
159 Ibid., p. 6.
160 "Directions for Erecting and Working the Newly Invented Steam Engines by Boulton and Watt" (1779). In: H. W. Dickinson and Rhys Jenkins, *James Watt and the Steam Engine* (Oxford, 1927), p. 394.

NOTES 141

161 "Directions for Working Rotative Engines" (ca. 1784), ibid., p. 400.
162 Walther von Dyck, *Georg von Reichenbach* (Munich, 1912) p. 5.
163 Dickinson, Jenkins, *James Watt and the Steam Engine,* p. 239, plates 40, 87, 90.
164 Denis Papin, *A New Digester, or engine for softening bones; containing the description of its make and use, etc.* (London, 1681).
Excerpts from this were published in *Acta Eruditorum Lipsiae* (1682):105-109.
165 Denis Papin, *Ars nova ad aquam ignis adminiculo efficacissime elevandam* (Frankfurt, 1707).
A report on this book was published in *Acta Eruditorum Lipsiae* (1707):420.
166 John Theophilus Desaguliers, *A Course of Experimental Philosophy,* 2 vols. (London, 1734/44), vol. 2, pp. 488-89.
167 British Patent (Old Series) No. 2302, 1799: Robert Delap, "Boilers for Steam Engines, etc." p. 3.
168 British Patent (Old Series) No. 2327, 1799: Matthew Murray, "Steam Engines" pp. 2-3.
169 Dickinson, Jenkins, *James Watt and the Steam Engine,* p. 239, plates 40, 90.
170 Charles Frederick Partington, *An Historical and Descriptive Account of the Steam Engine* (London, 1822), pp. 184-187; appendix, pp. 49, 59, 62; plate 8.
171 Conrad Matschoss, *Die Entwicklung der Dampfmaschine,* 2 vols., (Berlin, 1908), vol. 1, p. 598, figs. 465, 466.
172 This belief originates from Pierre de Latil's book *La pensée artificielle* (Paris, 1953); English under the title *Thinking by Machine: A Study of Cybernetics,* trans. Y. M. Golla (Boston, 1957); concerning the mill-hopper see pp. 116-118.
173 Herrad von Landsperg, *Hortus deliciarum,* ed. Joseph Walter (Strassbourg, 1952), pp. 82-83.
174 Agostino Ramelli, *Le diverse et artificiose machine* (Paris, 1588), for example part 3, no. 12.
175 Joh. Mathias Beyer, *Theatrum machinarum molinarum* (Leipzig, 1735), p. 23, plate 12.
176 A mill-hopper resembling that of Fig. 55 is found, for example, on the millstones of a late-19th century watermill exhibited in the Division of Agriculture, Deutsches Museum, Munich.
177 Karl Wilhelm Anton, *Encyclopädie für Müller* (Leipzig, 1871), p. 37.
178 Rex Wailes, "Some Windmill Fallacies," *Transactions of the Newcomen Society* 32(1959-60):97.
179 British Patent (Old Series) No. 615, 1745: Edmund Lee, "Self-regulating Wind Machine."
180 Gerard Doormann, *Patents for Inventions in the Netherlands during the 16th, 17th and 18th Centuries,* trans. Joh. Meijer (The Hague, 1942), p. 194.
181 British Patent (Old Series) No. 1041, 1773: John Barber, "Smelting Furnace, etc."
182 Rex Wailes, "Windmill Winding Gear," *Transactions of the Newcomen Society* 25(1945-47):31.

NOTES

183 William Fairbairn, *Treatise on Mills and Millwork*, 2 vols. (London, 1861), vol. 1, p. 277.
184 Rex Wailes, *The English Windmill* (London, 1954).
185 G. G. Schwahn, *Lehrbuch der praktischen Mühlenbaukunde* (Berlin, 1850), pt. 4, pp. 15-16.
186 D. W. Muggeridge, "The Windmills of Hanover," *Transactions of the Newcomen Society* 25(1945-47):201-202.
187 H. O. Clark, "Notes on French Windmills," *Transactions of the Newcomen Society* 9(1928-29):52 ("... automatic winding gear is unknown").
188 E. L. Burne, J. Russel, R. Wailes, "Windmill Sails," *Transactions of the Newcomen Society* 24(1943-45):151-154.
See also: Rex Wailes, *The English Windmill*, pp. 93-96.
189 Rex Wailes, "Notes on Windmill Drawings in Smeaton's Designs," *Transactions of the Newcomen Society* 28(1951-54):239-243.
190 British Patent (Old Series) No. 1588, 1787: Benjamin Heame, "Regulating the Sails of Mills."
191 British Patent (Old Series) No. 3041, 1807: William Cubitt, "Windmills."
192 Fairbairn, *Mills and Millwork*, vol. 1, p. 279; Moritz Rühlmann, *Allgemeine Maschinenlehre.*, 2nd ed. (Brunswick, 1875), vol. 1, p. 479.
193 Rex Wailes, "Some Windmill Fallacies," *Transactions of the Newcomen Society* 24 (1943-45): 151-154.
194 British Patent (Old Series) No. 1484, 1785: Robert Hilton, "Windmills."
195 British Patent (Old Series) No. 1628, 1787: Thomas Mead, "Regulator for Wind and Other Mills."
196 *Mechanics Magazine* 4 (1825):381.
197 Dickinson, Jenkins, *James Watt and the Steam Engine*, p. 222.
198 Theodor Beck, *Beiträge zur Geschichte des Maschinenbaues* (Berlin, 1899), p. 277.
Lynn White, Jr., *Medieval Technology and Social Change* (Oxford, 1962), pp. 115-116.
Bertrand Gille, *Engineers of the Renaissance*, trans. from the French (Cambridge, Mass., 1966), p. 111.
199 Huygens' description of the conical pendulum was originally published in his *Horologium oscillatorium* (Paris, 1673), reprinted in: *Oeuvres Complètes de Christiaan Huygens*, ed. Société Hollandaise des Sciences, (The Hague, 1934), vol. 18, pp. 360-365. A good summary of Huygens' work on the conical pendulum is given in Léopold Defossez, *Les Savants du 17e Siècle et la Mesure du Temps* (Lausanne, 1946), pp. 181-191. The article by V. C. Poor, "The Huygens Governor," *The American Mathematical Monthly* 32(1925):115-121 is purely an exercise in rational mechanics. The conical pendulum with which it deals is not identical with that of Huygens, nor is it a governor.
200 British Patent (Old Series) No. 1706, 1789: Stephen Hooper, "Regulating Wind and other Mills, etc."

201 E. L. Burne, J. Russell, R. Wailes, "Windmill Sails," *Transactions of the Newcomen Society* 24 (1943-45): 153.
Rex Wailes, *The English Windmill* (London, 1954), p. 94.
202 William Fairbairn, *Treatise on Mills and Millwork* (London, 1861), vol. 1, pp. v-x.
Friedrich Klemm, *A History of Western Technology*, trans. D. W. Singer, (Cambridge, Mass., 1964), pp. 234-240.
203 James Watt, Letter to Matthew Boulton, Birmingham, Sept. 5, 1783. In: James Patrick Muirhead, *The Origin and Progress of the Mechanical Inventions of James Watt* (London, 1854), vol. 2, p. 177.
204 John Bourne, *Treatise on the Steam Engine* (London, 1849), p. 9.
205 A. Stowers, "The Development of the Atmospheric Steam Engine after Newcomen's Death in 1729," *Transactions of the Newcomen Society* 35(1962-63):91.
206 Dickinson, Jenkins, *James Watt and the Steam Engine*, pp. 183-184.
207 Ibid., pp. 162-164.
208 Cyrill T. G. Boucher, *John Rennie, 1761-1821* (Manchester, 1963), pp. 10-11. See also note 152.
209 Dickinson, Jenkins, *James Watt and the Steam Engine*, pp. 164-167.
210 The following account of the invention of James Watt's centrifugal governor is based upon the chapter "The Governor and the Throttle Valve" in Dickinson, Jenkins, *James Watt and the Steam Engine*, pp. 220-223.
211 Boulton and Watt Collection, ibid., p. 220.
212 John Robison, *Steam and Steam Engines*, with an appendix by James Watt, in: *A System of Mechanical Philosophy*, 5 vols., (Edinburgh, 1822), vol. 2, p. 155.
213 Some authors claim that the centrifugal governor is covered by James Watt's Patent No. 1432 of 1884 ("Certain new Improvements upon Fire and Steam Engines and upon Machines worked or moved by the same"). The 6th part of this patent ("regulating valves"), however, deals with the valve gear, not with the governor.
214 Dickinson, Jenkins, *James Watt and the Steam Engine*, pp. 359-360.
215 Ibid., p. 221.
216 British Patent (Old Series) No. 2782, 1804: John Bywater, "Windmills."
217 Thomas Young, *A Course of Lectures on Natural Philosophy and the Mechanical Arts*, 2 vols. (London, 1807), vol. 1, p. 47, 190, 233.
218 J. N. P. Hachette, *Traité élémentaire des machines* (Paris, 1811), p. 139, plate 12.
219 Robertson Buchanan, *Practical Essays on Millwork and other Machinery*, 2 vols., 2nd ed. (London, 1823), vol. 2, pp. 477-496, plates 13-15.
220 Olinthus Gregory, *A Treatise of Mechanics, Theoretical, Practical, and Descriptive*, 3rd ed., 3 vols. (London, 1815), vol. 2, pp. 204-208.
221 J. A. Borgnis, *Traité complet de Mécanique appliquée aux arts*, 9 vols., (Paris, 1818-1821), vol. 1, p. 378.
222 Oliver Evans, *Manuel de l'ingénieur Mécanicien Constructeur de Machines à Vapeur*, trans. I. Doolittle, (Paris, 1821), pp. 175-178.

223 Oliver Evans, *The Abortion of the Young Steam Engineer's Guide* (Philadelphia, 1805).
224 Greville and Dorothy Bathe, *Oliver Evans: A Chronicle of Early American Engineering* (Philadelphia, 1935), p. 286.
225 Christoph Bernoulli, *Anfangsgründe der Dampfmaschinenlehre* (Basel, 1824), pp. 172-75.
226 Karl Christian von Langsdorf, *Ausführliches System der Maschinenkunde* (Heidelberg, 1826), vol. 1, pp. 496-99.
227 Jean-Victor Poncelet, *Cours de mécanique appliqué aux machines* (1826), ed. X. Kretz (Paris, 1874), part 2.
228 John Farey, *A Treatise on the Steam Engine* (London, 1827), pp. 465-467.
229 Thomas Tredgold, *The Steam Engine: Its Invention and its Progressive Improvement* (1827), 2nd. ed., (London, 1838), vol. 1, pp. 263-264, appendix p. 213.
230 Zachariah Allen, *The Science of Mechanics* (Providence, 1829), pp. 273-76.
231 Wilhelm Hort, "Die Entwicklung des Problems der stetigen Kraftmaschinenregelung," *Zeitschrift für Mathematik und Physik* 50 (1904):233-279.
232 Charles Frederick Partington, *An Historical and Descriptive Account of the Steam Engine* (London, 1822), pp. 135, 157.
233 Robert Stuart, *A Descriptive History of the Steam Engine* (London, 1824), pp. 133-134.
234 Elijah Galloway, *History of the Steam Engine* (London, 1828), p. 36.
235 François Arago, *Oeuvres complètes,* ed. J.-A. Barral (Paris, 1865), vol. 5, pp. 1-116.
236 Matschoss, *Entwicklung der Dampfmaschine,* vol. 1. 221
237 M. R. de Prony, *Nouvelle architecture hydraulique,* 2 vols., (Paris, 1790/96), vol. 1, pp. 571-72; vol. 2, pp. 49-52.
238 Ibid., vol. 2, pp. 50 and 52.
239 Borgnis, *Traité complet de Mécanique,* vol. 1, 375-78.
240 Partington, *Steam Engine,* p. 135.
241 Tredgold, *Steam Engine,* p. 264.
242 Poncelet, *Cours de mécanique,* vol. 2, pp. 75-81.
243 Willibald Trinks, *Governors and the Governing of Prime Movers* (New York, 1919), p. 143-144.
244 Alfred Chapuis, *A.-L. Breguet pendant la Révolution française* (Neuchâtel, 1953), p. 94.
245 Hans von Bertele, "Selbstoptimierende, mechanische Systeme vor 150 Jahren," *Die BASF* 14(1964):124-127.
246 Adam Smith, *An Inquiry into the Nature and Causes of the Wealth of Nations* (1776), ed. Edwin Cannan (New York, 1937), for example pp. 56-59. For a detailed discussion see Otto Mayr, "The Feedback Concept in Economic Thought and Technology in 18th-Century Britain," *Technology and Culture,* Vol. 12, No. 1 (1971).
247 David Hume, "Of the Balance of Trade," *Political Discourses* (Edinburgh, 1752). Reprinted: Eugene Rotwein, *David Hume: Writings on Economics* (Edinburgh, 1955), p. 62-63.

248 Johann Georg Krünitz, *Ökonomisch-technologische Enzyklopädie*, 242 vols., (Berlin, 1773-1858), vol. 121 (1812).
249 Abraham Rees, *The Cyclopaedia*, 39 vols. (London, 1819), vol. 29.
250 *Encyclopaedia Britannica*, 6th ed., 20 vols., (Edinburgh, 1823), vol. 17.
251 *Dictionnaire Technologique*, 22 vols., (Paris, 1822-1835), vol. 18 (1831).
252 J.-A. Borgnis, *Traité complet de Mécanique*, vol. 1, pp. 324-383.
253 J.-A. Borgnis, *Dictionnaire de Mécanique appliquée aux arts* (Paris, 1823), s.v. régulateur.
254 For a survey of the history of control engineering from the early 19th to the mid-20th century, see Klaus Rörentrop, *Zur Entwicklung der Regelungstechnik*, Dissertation, University of Erlangen-Nürnberg (Erlangen, 1969).
255 George Biddell Airy, "On the Regulator of the Clock-Work for effecting uniform Movement of Equatorials," *Memoirs of the Royal Astronomical Society* 11(1840):249-267.
Idem, supplement to the above paper, *Memoirs of the Royal Astronomical Society* 20(1851):115-119.
256 James Clerk Maxwell, "On Governors," *Proceedings of the Royal Society* 16(1867/68):270-283. Reprinted: Richard Bellman and Robert Kalaba, editors, *Mathematical Trends in Control Theory* (New York, 1964), pp. 3-17.
257 Ivan A. Wischnegradski, "Sur la théorie générale des régulateurs," *Comptes rendus... de l'Académie des sciences* 83 (1876):318-321.
Idem, "Mémoire sur la théorie générale des régulateurs," *Revue universelle des Mines* 4(1878):1-38 and 5(1879): 192-227.
258 E. J. Routh, *A Treatise on the Stability of a Given State of Motion* (London, 1877), esp. pp. 42-44.
259 A. M. Lyapunov, *On the General Problem of Stability of Motion*, in Russian, (Kharkov, 1892).
260 A. Stodola, "Über die Regulierung der Turbinen," *Schweizerische Bauzeitung* 22(1893):113-17, 121-22, 126-28, 134-35 and 23 (1894):108-12, 115-17.
261 A. Hurwitz, "Über die Bedingungen, unter welchen eine Gleichung nur Wurzeln mit negativen reellen Teilen besitzt," *Mathematische Annalen* 46(1895):273-284.
Reprinted: Bellman and Kalaba, *Mathematical Trends in Control Theory*, pp. 70-82.
262 Max Tolle, "Beiträge zur Beurteilung der Zentrifugalpendelregulatoren," *Zeitschrift des Vereines deutscher Ingenieure* 39(1895): 735-41, 776-79, 1492-98, 1543-51; 40(1896):1424-28, 1451-55.
263 K. Küpfmüller, "Über die Dynamik der selbsttätigen Verstärkungsregler," *Elektrische Nachrichten-Technik* 5(1928):459-467.
264 H. Nyquist, "Regeneration Theory," *Bell System Technical Journal* 11(1932):126-147.
Reprinted: Bellman and Kalaba, *Mathematical Trends in Control Theory*, pp. 83-105. Retrospective remarks by H. Nyquist, "The Regeneration Theory," are published in Rufus Oldenburger, editor, *Frequency Response* (New York, 1956), p. 3.

265 H. L. Hazen, "Theory of Servomechanisms," *Journal of the Franklin Institute* 218(1934):279-331.
266 H. S. Black, "Stabilized Feed-Back Amplifiers," *Bell System Technical Journal* 13(1934):1-18.
267 R. C. Oldenbourg and H. Sartorius, *Dynamik selbsttätiger Regelungen* (Munich and Berlin, 1944), English translation: *The Dynamics of Automatic Controls,* trans. and ed. H. L. Mason (New York, 1948).
268 LeRoy MacColl, *Fundamental Theory of Servomechanisms* (Princeton, N.J., 1945).

Picture Credits

(Figure numbers in boldface)

1: Matschoss, Conrad, *Die Entwicklung der Dampfmaschine* (Berlin 1908), Vol. 1, Fig. 107.

2, 54: Dickinson, H. W., Jenkins, Rhys, *James Watt and the Steam Engine* (Oxford 1927), Tables 81 and 90.

5: Diels, Hermann, *Antike Technik*, 3rd ed. (Leipzig 1924), Fig. 71.

7, 10, 11, 13, 14, 15: Schmidt, W., ed., *Heron von Alexandria: Druckwerke und Automatentheater,* Greek and German (Leipzig 1899), Figs. 122, 21, 75, 74, 20, 26.

18, 19, 20: Wiedemann, E., Hauser, F., "Uhr des Archimedes und zwei andere Vorrichtungen," *Nova Acta* Abh. d. Kaiserl. Leop.-Carol. Dtsch. Akad. d. Naturf., Vol. 103/2 (Halle a/Saale 1918), Figs. 1, 1a, (1).

21, 22, 23, 24, 25, 26, 27: Wiedemann, E., "Über die Uhren im Bereich der islamischen Kultur," *Nova Acta* Abh. d. Kaiserl. Leop.-Carol. Dtsch. Akad. d. Naturf., Vol. 100/5 (Halle a/Saale 1915), Figs. 19, 21, 22, 38, 93, 98, 131.

28, 29, 30, 31, 32, 33, 34: Wiedemann, E., "Über Trinkgefässe und Tafelaufsätze nach al-Ğazarī und den Benū Mūsā," *Der Islam,* Vol. 8 (1918), Figs. A, B, D, E, Ea, F, Fa.

35: Hauser, F., "Über das kitāb al ḥijal ... der Benū Mūsā," *Abh. z. Gesch. d. Naturwiss. u. d. Medizin,* Vol. 1 (1922), Fig. 95.

148 PICTURE CREDITS

38, 39: Reproduced in *Annals of Science,* Vol. 6 (1948), pp. 40/41, Figs. 1, 2.

40: Monconys, Balthasar de, *Journal des Voyages* (Lyon 1665-66), Section 2, p. 42, Fig. 5.

42: Commandino, F., ed., *Heron de Alexandria, Spiritualium liber* (Urbino 1575), p. 69.

43: *Repertory of Arts and Manufactures* (London 1796) Vol. 5, Table III.

45, 46: *Dinglers Polytechnisches Journal* (1828), vol. 29, Table III.

47: Popow, E. P., *Einführung in die Regelungs- und Steuerungstechnik* (Berlin 1964), Fig. 1.12.

English Patents of Inventions (Old Series):

48: S. T. Wood, No. 1447 (1784), Fig. 7.

51: R. Delap, No. 2302 (1799), Figs. 1, 8.

53: M. Murray, No. 2327 (1799), Fig. 1.

58: E. Lee, No. 615 (1745).

61: R. Hilton, No. 1484 (1785), Fig. 5.

62, 64: T. Mead, No. 1628 (1787), Figs. 20, 26-46.

63, 66: S. Hooper, No. 1706 (1789), Figs. 3, 1.

49: *Acta eruditorum* (Leipzig 1682), p. 106.

55: Anton, K. W. *Encyclopädie für Müller* (Leipzig 1871), Vol. 5, Fig. 1.

57: Fairbairn, W., *Mills and Millwork* (London 1861), Section 1, Figs. 174, 176.

67: Bourne, John, *The Steam Engine* (London 1846), p. 9.

68: Prony, R., *Nouvelle architecture hydraulique* (Paris 1796), Table 26.

69: Chapuis, A. *Breguet pendant la Révolution française* (Neuchâtel 1953), p. 95.

70: *Die BASF* (1964), vol. 14, p. 127.

Index

Airy, G. B., 131
Albion Mill, 110, 111
Alfonso X "the Wise" of Castille, 38
Ampère, A. -M., 2
Antikythera, 26
Athanor, Drebbel's, 58, 59, 61, 62
Automata, Heron's, 19
Automatic control, theory of, 131-132

Baille-blé. *See* Mill-hopper
Banū Mūsā, 18, 40, 41-46
Barber, John, 95, 97
Becher, Johann Joachim, 66, 67
Bīrūnī, al-, 46
Black, H. S., 132
Block-diagram notation, 4-5
Bonnemain, 70, 71, 72-75
Borgnis, J.-A., 114, 117, 130, 131
Boulton, Matthew, 110
Boulton & Watt, 80, 81, 87, 110, 111
Boyle, Robert, 56
Brahmah, J., 76
Breguet, Abraham-Louis, 119-124
Brindley, James, 77, 128
Brunton, William, 89
Buchanan, Robertson, 114

Cataract, Boulton & Watt's, 110
   Smeaton's, 110
Centrifugal governor, 1-4, 113, 114, 115, 125
   Huygens', 102
   notation for, 5
   Watt's, 1, 204, 109-112, 113
Centrifugal pendulum, 100, 101, 102
   Huygens', 102
   Mead's, 100
Chemical furnace, Drebbel's, 63
Clock of Gaza, 28
Closed-loop control. *See* Feedback control
Commandino, Federigo, 47, 48, 65, 127
Comparator, 8
Cubitt, William, 94, 97, 98
Cumming, A., 76
Cybernetics, 2, 132

da Vinci, Leonardo, 18, 38, 46, 47
de Béthencourt, Augustin, 115
de Conti, Prince Louis-François de Bourbon, 67, 68
Delap, Robert, 83-86, 128
de Monconys, Balthasar, 57, 62, 67
de Peiresc, N. C. Fabri, 55, 56, 57
de Prony, M. R., 115
de Romas, Jacques, 68
Desaguliers, Jean-Théophile, 82
Diels, Hermann, 11, 15
Drebbel, Cornelis, 55-65

Evans, Oliver, 114

Fan-tail, 93-96

INDEX

Feedback control, 1, 129
  closed-loop, 7, 8, 67, 78, 93, 129
  definition of, 2, 7, 78
  evolution of, 125-132
  general concept of, 129
  identification of, 6-8
  invention of, 63
  manual, 7
  on mills, 90-108
  negative, 7
  on-off, 18, 44
  self-regulating systems, 6, 98
  servomechanisms, 6
  watch synchronization, 119-124
Float-valve regulators, 15, 16, 25, 40, 47, 48, 51, 52, 76-81, 87, 125, 126, 127
  of Banū Mūsā, 40, 42, 43, 44, 45, 65
  of Brindley, 77
  drinking-straw (Chinese), 50, 51
  etymology of, 78
  of Heron of Alexandria, 15, 21, 42, 65
  of al-Jazari, 33, 34, 35
  of Ktesibios, 15, 16, 20, 25, 33
  of Polzunov, 77, 78
  of Pseudo-Archimedes, 32, 33
  of al-Sā'ātī, 36, 37
  in water clocks, 28, 38-39
Flywheel, 102
Furnace, Drebbel's, 57, 58-64, 65

Gaza, clock of, 28
Governor, etymology of, 2

Hachette, J. N. P., 113
Hazen, H. L., 132
Heame, Benjamin, 97
Henry, William, 69, 70
Heron of Alexandria, 15, 18, 19-25, 26, 27, 40, 41, 42, 46, 47, 48, 64, 65, 76, 78, 127
Heron of Byzantium (Heron the Younger), 40
Hilton, Robert, 99
Hooke, Robert, 47, 56, 66
Hooper, Stephen, 102, 103, 106, 107
Hume, David, 129
Hurwitz, A., 131
Huygens, Christiaan, 102

ibn Ishāq, Hunayn, 41
ibn Lūqā, Qustā, 41
ibn Qurra, Thābit, 41
Incubator, Réaumur's, 67
  Drebbel's, 58, 59, 61, 62, 63
Integral control, 117

Jazari, al-, 28, 29, 32, 33-34, 35, 38, 39

Kitāb al-Hiyal (Banū Mūsā's), 41, 42
Ktesibios, 11-15, 20, 25, 31, 33
Kuffler, Abraham, 56
Kuffler, Augustus, 57
Kuffler, Johan Sibertus, 57, 60, 63
Küpfmüller, K., 132

Landsperg, Herrad von, 90, 91
Lap engine, 111
Lee, Edmund, 93-96
Leurechon, Jean, 18, 47
Liberalism, 128, 129
Lift-tenters, 99
  Hooper's, 103
  Mead's, 100, 101, 110
Lyapunov, A. M., 131

MacColl, LeRoy, 132
Marriotte's bottle, 38
Maxwell, J. C., 131
Mead, Thomas, 100, 101, 102, 103-105, 107
Mechanica (Heron's), 41
Meikle, Andrew, 93, 96, 97, 110
Mercantilism, 128, 129
Mill-hopper, 90, 91, 92, 93
Millwrights, 107, 108
Modérateur, 115, 129, 130
Murray, Matthew, 86, 87, 128

Nyquist, H., 132

Offset, proportional, 3
Oil lamp, of Banū Mūsā, 44, 45, 46
  of Hooke, Robert, 66
  of Philon, 16, 17-18, 23, 25, 42, 44, 47
Oldenbourg, R. C., 132
Oldenburg, Henry, 60
On-off control. *See* Feedback control

Papin, Denis, 82-84
Patent-sails, Cubitt's, 97
Pendule sympathique, Breguet's, 119-124
Périer brothers, 115, 116-118, 128
Philon of Alexandria, 16-18, 23, 25, 41, 44, 47
Pneumatica, Heron's, 19-21, 25-27, 40, 42, 46-48, 64, 65, 76, 78, 127
Pneumatica, Philon's, 41
Polzunov, Ivan Ivanovich, 77, 78, 128
Pressure regulators, 82-89
  Boulton & Watt's, 87, 88, 89
  Brunton's, 89

INDEX 151

Pressure regulators (*continued*)
  Delap's, 83, 84, 85, 86
  Murray's, 86, 87
  on-off control, 86
  Papin's, 82, 83
Proportional control, 87, 117
Pseudo-Archimedes, 28-33, 36
Pseudo-Heron, 40

Ramelli, Agostino, 47, 90
Réaumur, René-Antoine Ferchault de, 67, 68, 71
Regulating systems. *See* Feedback control
Regulators, float-valve. *See* Float-valve regulators
  governors, use of term, 129-131
  pressure. *See* Pressure regulators
  speed. *See* Speed regulators
  temperature. *See* Temperature regulators
Reichenbach, Georg von, 80-81, 113
Rennie, John, 110, 113
Ridwān al-Khurasānī. *See* Sā'ātī, Ibn al-
Roller reefing sails, Hooper's, 106, 107
Routh, E. J., 131

Sā'ātī, Ibn al-, (Ridwān al-Khurasānī), 28, 29, 32, 35, 36, 38
Safety valve, Desaguliers', 82
  Papin's, 82
Salmon, William, 76, 125
Sampled-data control systems, 123
Sartorius, H., 132
Schwenter, Daniel, 47, 66
Self-regulation. *See* Feedback control
Sensing devices, rotary speed, 99-101, 102
  Hilton's, 99, 100
Sensing element, 8
Sentinel Register, Henry's, 69-70
Servo system, 95. *See also* Feedback control
Smeaton, John, 110
Smith, Adam, 128
Soulas, Achille Elie Joseph, 51
Southern, John, 111
South-pointing chariot, 49
Speed regulators, 102, 103-106, 107
  Hooper's, 106, 107
  Mead's, 102, 103-105, 107
  Périer brothers', 115, 116-117

Spring sail, Meikle's, 96, 97
Steam-engine
  rotating, 110
  Southern's "Lap," 111
  speed regulation of, 109-118
  Watt's, 2-4, 109-112
Stodola, A., 131
Syphon, floating, Heron's, 25

Temperature feelers, Bonnemain's, 71, 72-74
  de Romas's, 68
  Henry's, 69
  Réaumur's, 68
Temperature regulators, 55-75
  of Becher, 66, 67
  of Bonnemain, 70, 71, 72-75
  of Drebbel, 57, 58-65
  of Henry, 69, 70
  of Réaumur, 67, 68
Temporary hour, 13
Thermostat, origin of word, 74, 75
Throttling valve, 112
Tolle, S. M., 131

Ure, Andrew, 74, 75

Valla, Giorgio, 46
Vitruvius, 15, 33
Vyshnegradskii, I. I., 131

Water clock, 25, 27, 28, 38-39
  of al-Jazarī, 33, 34, 35, 38
  of Ktesibios, 11-15, 25, 31, 33
  of Pseudo-Archimedes, 28-30, 31, 32, 33
  of al-Sā'ātī, 36, 37, 38
Watt, James, 1, 2-4, 80, 81, 102, 108, 113, 128
Wiener, Norbert, 2, 7
Windmill sails, self-regulating, Barber's, 97
  Cubitt's, 94, 97
  Heame's, 97
  Lee's, 96, 97
  Meikle's, 96
Wood, Sutton Thomas, 78, 79, 128
Wren, Sir Christopher, 56, 60

Young, Thomas, 113